Inside SCIENCE

Edited by John Allen

BBC Books

Published to accompany
the series of programmes
produced in consultation with
the BBC Educational Broadcasting Council.

Published by BBC Books,
a division of BBC Enterprises Limited,
Woodlands, 80 Wood Lane, London W12 0TT
First published 1988

© The Authors 1988

ISBN 0 563 21410 4

Set in Bembo 10 on 11 and Helvetica 9 on 10
Photoset, printed and bound in Great Britain by
Redwood Burn Limited, Trowbridge, Wiltshire
Cover printed by Fletchers of Norwich

Book and cover design by Gail Hurstfield
Illustrations chapters 5 and 10 by John Gilkes
All other illustrations by Lizzie Kelsall

Contents

	Acknowledgements	4
	Introduction	5
1	Nuclear energy *The atomic powerhouse*	8
2	The energy 'crisis' *Power for our future*	24
3	Test tube babies *Creating new life*	38
4	Pollution *The big clean-up*	52
5	Lasers *The light fantastic*	66
6	Plastic *Our flexible friend*	82
7	Healthy eating *The facts about food*	96
8	The changing Earth *Eruptions and earthquakes*	110
9	Building roads *The way ahead*	122
10	Artificial intelligence *The electronic brain?*	134
	List of contributors	150
	Index and glossary	152

Acknowledgements

The publishers would like to thank the following for permission to reproduce illustrations based on diagrams which have appeared in their publications:

JOHN WILEY AND SONS INC, USA (H. A. Sorenson, *Energy Conversion Systems*, 1983) p. 26 *above*; PERGAMON PRESS (J. C. McVeigh, *Energy Around the World*) p. 26 *below*; SWEDISH NATIONAL ENVIRONMENTAL PROTECTION BOARD (*Acidification: A Boundless Threat to Our Environment*) pp. 53 and 54; OPEN UNIVERSITY (Course S101, Unit 28) p. 111; MACMILLAN PUBLISHING COMPANY (F. Sawkins, Chase, Darby and Rapp, *The Evolving Earth*, 1978) pp. 114 and 115.

Picture credits

WIND ENERGY GROUP p. 27 *above*; NATIONAL ENGINEERING LABORATORY/D.T.I. p. 27 *below*; CHAMBRE DE COMMERCE ET D'INDUSTRIE DE SAINT MALO p. 28; PHIL MATSON p. 41 *above*; VIRGINIA BOLTON p. 41 *below*; JOHN PARSONS p. 44; FRIENDS OF THE EARTH p. 56 *left*; FRIENDS OF THE EARTH/CHRIS ROSE p. 56 *right*; BARNABY'S PICTURE LIBRARY p. 56 *below*; NIGEL GRAHAM p. 59; BXL PLASTICS LTD p. 91; ALFRED MCALPINE PLC p. 124; ALFRED MCALPINE PLC p. 131 *left*; TARMAC/F. R. LOGAN PHOTOGRAPHY p. 131 *right*, TONY TIMMINGTON/BBC cover.

Introduction

Some of my earliest memories are of taking things apart to find out how they worked (and only occasionally getting them back together again!), so 'how' and 'why' have always been important questions to me. My fascination with science meant that making the Radio 4 series of *Inside Science* was an interesting challenge, firstly to find out the answers to questions like 'how does a hologram work?' and then, with Paul Heiney's help, to explain them to the listener. In the process we have tried to show that science is not only important, it's also interesting.

Our approach to the series was to take ten scientific topics that regularly find their way into the headlines and, using an expert as a guide, to delve into the science behind them. In this book those same experts each contribute a chapter, giving background information in straightforward language and putting the science involved into the context of everyday life. It works as a kind of briefing on many of the issues that emerge from present-day science, a briefing that doesn't make all sorts of assumptions about the reader's level of background knowledge.

Both the series and the book aim to provide the information needed to understand the arguments on issues such as whether we should have nuclear power stations, or the whys and wherefores of human embryo research. Because the political implications of science are widely discussed elsewhere, any conclusions drawn in this book are based on science alone, although of course its political and ethical consequences cannot be ignored.

Apart from giving us a more informed view of these major issues, an understanding of science can also demystify much of what surrounds us in the modern world. There's no doubt that our lives have become totally dependent on the work of the scientist. Try to imagine life without cars, electric lighting, ballpoint pens, the telephone, vacuum cleaners or even the much-blamed computers. You often hear people talking about 'the good old days' with nostalgia, but were they really better? It's hard to believe that anyone would willingly give up the convenience of electric lighting or forego their washing-machines and televisions.

There is much here for the general reader but *Inside Science* also covers many of the subjects found in GCSE syllabuses for physics, chemistry and biology; examples are light, atoms and molecules, and reproduction. In *Inside Science* the context explains and underlines the importance of these basic concepts. These days science and engineering are closely related, so technology also plays a part throughout the book and especially in specific areas such as the chapter on artificial intelligence. As you read the text you will notice that certain words have been printed in bold type. This simply means that they are explained in the glossary on page 152.

Although *Inside Science* was not written as a textbook, it can be used, together with the radio programmes, to extend the reader's scientific knowledge within a more formal educational framework. And that's where the *Inside Science* software pack for the BBC micro comes in. It's available from BBC Software and contains four programs to amplify specific points. For instance there is an adventure game in which the user takes on the role of a Pollution Control Officer trying to locate and control a source of pollution in a river. There's also a game of noughts and crosses in which the computer plays against the user and uses a neural network to learn how to win – if you use a particular winning strategy the computer will copy it! Other programs include a simulation of the operation of a nuclear power station, and a package on selecting suitable man-made or natural materials to use in manufacturing a car or designing a kitchen.

If you would like to use *Inside Science* as part of a course there is a free factsheet giving lots of ideas on how it can be used. Simply send a large stamped addressed envelope to: Inside Science Factsheet, BBC Education, London W5 2PA.

You may of course be interested in the City and Guilds of London Institute Inside Science course. Further details about this course, which leads to a computer-assessed certificate, are available from: Inside Science (7760), Sales Section, City and Guilds of London Institute, 76 Portland Place, London W1N 4AA.

We all learnt a great deal from the making of *Inside Science*. I hope you find the process of investigation as exciting as we did.

John Allen
Producer, *Inside Science*

1 NUCLEAR ENERGY
The atomic powerhouse

Helen ApSimon and Tony Goddard

> **This chapter investigates:**
> - what happened at Chernobyl
> - nuclear power
> - splitting atoms
> - nuclear power reactors
> - radiation and its effects
> - measuring radioactivity
> - the nuclear fuel cycle
> - radiation in the environment
> - nuclear power in the future

What happened at Chernobyl?

At 1.23 a.m. local time on 26 April 1986 a series of explosions shook the Number 4 reactor at the Chernobyl nuclear power station in the Ukraine. The energy of these explosions was minuscule compared with a nuclear bomb, but something had occurred which many scientists and engineers believed was never likely to happen – the top of the reactor had been torn open, exposing a burning **core** which sprayed radioactive debris into the air. It took heroic efforts over the next ten days to control this release, during which time several thousand tonnes of boron, lead, dolomite and sand were dropped from helicopters to prevent any further nuclear reactions and effectively seal the core before its final covering in concrete.

Some of the first casualties were firemen, many of whom suffered severe **radiation** burns fighting secondary fires caused by hot fuel fragments falling on flammable roof materials on other reactor buildings. Soviet scientists, reporting in 1986 at the International Atomic Energy Agency, indicated that 31 people had died and 135 000 people had been evacuated from within a 30 km radius.

Immediate deaths and injuries were not the only effects of the accident, however. Enormous sums of money have been spent since on decontamination and protection of water supplies. Some future cancers will be due to Chernobyl although they will not be detectable amongst the cancers caused by other factors. The **radioactivity** also penetrated far beyond the USSR across Europe, giving rise to further problems. Areas where heavy rainfall caused concentrated patches of ground contamination were particularly badly affected. In Britain this occurred over North Wales, the Lake District and Scotland, and has resulted in sales of sheep being restricted from some of these areas over a long period of time.

Chernobyl was therefore a very major accident, releasing a substantial proportion of the radioactivity in the core to the environment. It resulted from violations of safety regulations and design flaws in this type of reactor, and emphasised the need for great care in the design, construction and operation of nuclear reactors. (The automatic mechanisms which shut down reactors safely if any irregularity or failure occurs are obviously vital.) It has also focused attention worldwide on many questions concerning nuclear power and its safety.

What is nuclear power?

Four centuries before Christ, Democritus postulated that the physical material of the universe was made up of **atoms**. Over the years we have increasingly used the chemical energy stored in combinations of atoms. For example, the burning of fossil fuels, such as coal and oil, releases chemical energy. However the last century has seen rapid advances in our knowledge of the structure of atoms, and in particular the **nuclei** at their centres. It is the energy stored in the nuclei of a particular type of atom called uranium-235 (or ^{235}U for short) that is the source of energy for nuclear power. And this energy is released when such nuclei are split.

An atom is made up of a nucleus containing two kinds of particles: **protons** (each one carries a positive electric charge) and **neutrons** (no electric charge).

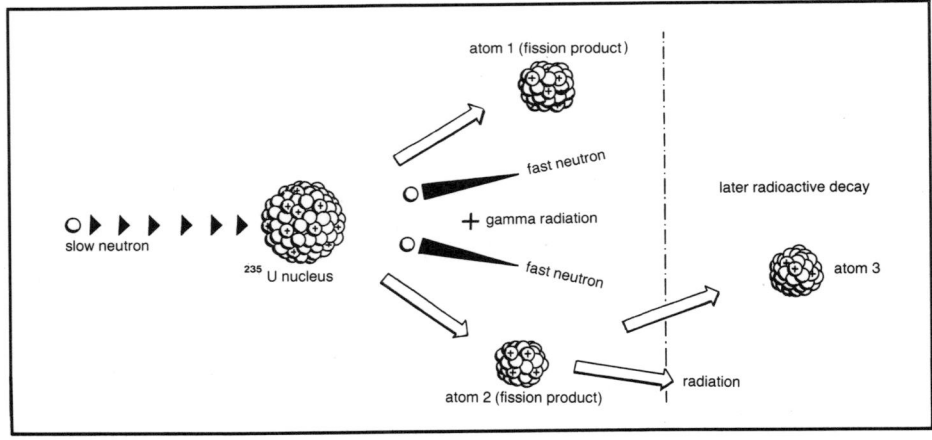

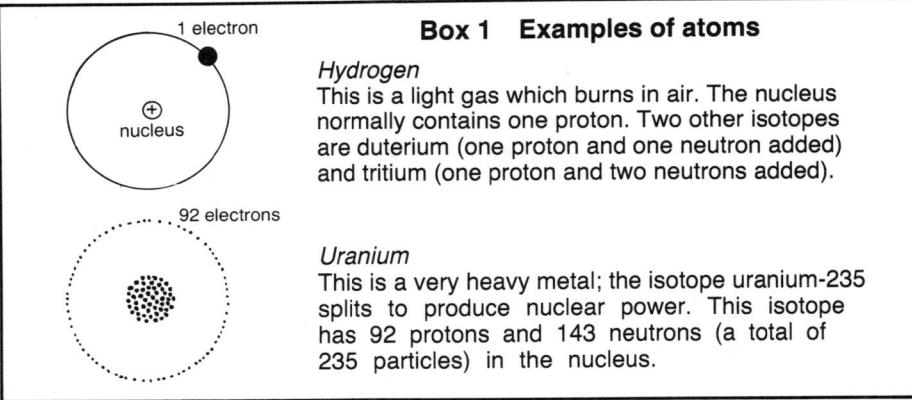

Box 1 Examples of atoms

Hydrogen
This is a light gas which burns in air. The nucleus normally contains one proton. Two other isotopes are duterium (one proton and one neutron added) and tritium (one proton and two neutrons added).

Uranium
This is a very heavy metal; the isotope uranium-235 splits to produce nuclear power. This isotope has 92 protons and 143 neutrons (a total of 235 particles) in the nucleus.

Figure 1, *top: Fission of ^{235}U nucleus;* **Figure 2,** *above: Examples of atoms – a gas (hydrogen) and a metal (uranium)*

In orbit around the nucleus are the **electrons**, which carry negative electric charges to balance the positive charges on the protons. Electrons are much lighter than a proton or a neutron – only about one two-thousandth of their mass.

The number of electrons always equals the number of protons, and is known as the **atomic number**. This number tells us about the chemical character of the atom.

Atoms with the same atomic number and chemical behaviour can have different numbers of neutrons in the nucleus and hence have different atomic masses. This leads to different **isotopes** (or forms) of the same **element**. So another isotope of

NUCLEAR ENERGY

uranium is uranium-238 (^{238}U) which has 92 protons but 146 neutrons (instead of 143 as in ^{235}U), giving a total of 238 particles in the nucleus, and an **atomic mass** of 238.

Splitting atoms and the chain reaction

As mentioned earlier, nuclear energy can be released when the nucleus of an atom splits. This splitting process is known as nuclear **fission**. To take an example, the nucleus of an atom of ^{235}U can be made to split by bombarding it with a neutron particle. (This is easier if the neutron is not moving too fast.) The nucleus splits into two smaller parts, forming the nuclei of new atoms called fission fragments or **fission products**. The nucleus does not always split in exactly the same way and so a variety of different fission product atoms can result.

The combined mass of the new atoms is less than that of the original nucleus, and since as Einstein declared, mass can be converted into energy, there is energy to spare. How does this energy appear and how can we use it? The answer is that the energy is mostly given to the fission fragments, which move away at high speed, producing **kinetic energy**. If the original atom is part of a bar of uranium this energy is given up as the fragments slow down by colliding with other atoms, and the bar heats up.

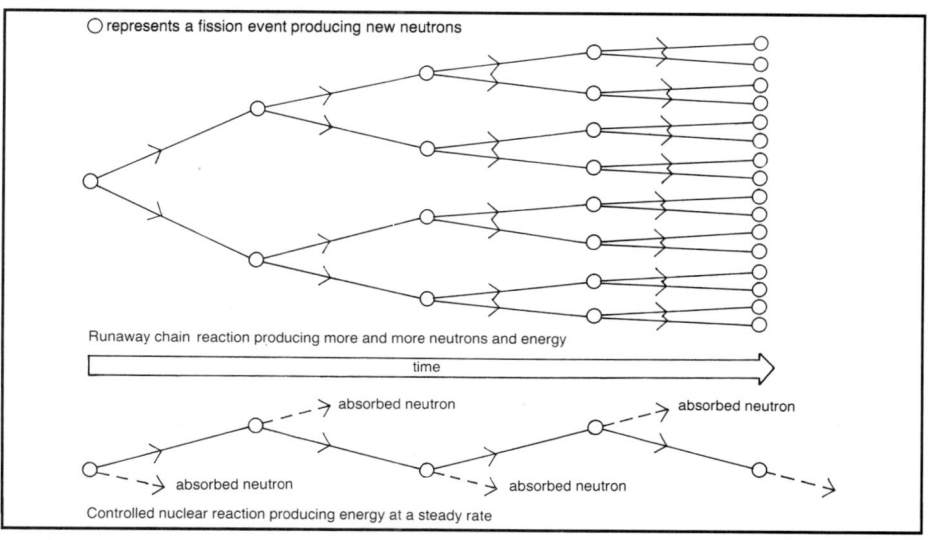

Figure 3, *below: Chain reactions*

New fast-moving neutrons are also thrown out during fission. Each time a fission occurs, two (or sometimes three) neutrons are ejected from the nucleus as it splits.

Because each fission of a ^{235}U nucleus produces at least two new neutrons, these new neutrons can be slowed down and used to split further ^{235}U nuclei. If all the new neutrons produced go on to cause more fissions then the number of fissions keeps on multiplying in a **chain reaction**, with a runaway increase in energy produced. This is what happens in a nuclear explosion.

In a nuclear power station the chain reaction has to be controlled, so that on average only one neutron from each fission goes on to cause further fission – the rest of the neutrons being absorbed and lost. The fission energy can then be released at a steady rate and carried away by a coolant flowing over the hot fuel. Nuclear power plants are designed for this controlled release of energy, allowing no possibility of a runaway chain reaction.

Nuclear power reactors

A nuclear power reactor is constructed like a 'nuclear boiler' to extract nuclear energy and generate electricity from it.

- The fuel is uranium – first mined as uranium ore, then processed and packed into metal cans to form **fuel elements**. A large number of fuel elements are arranged together in the reactor core.

- The **control rods** contain boron which absorbs neutrons very effectively. These act as the 'accelerator' and 'brake'. When they are raised out of the core the number of unabsorbed neutrons (and, therefore, the fission power) is increased. When the rods are placed deeper in the core the number of neutrons (and fission power) is reduced.

- The **moderator** is a substance such as graphite which slows down neutrons so that they can cause further fission. Neighbouring fuel elements are separated by the moderator.

NUCLEAR ENERGY

- The **coolant** circulates through the reactor core. The core becomes very hot from the nuclear energy produced within the fuel elements, and so the coolant carries this heat away along closed pipes to a heat exchanger. Having given up its heat, the coolant is recirculated through the reactor core in a closed loop.

- The heat exchanger generates steam which is used to drive turbines in the turbine hall, to generate electricity, as in other types of power station.

- Surrounding the core, a **shield** of thick concrete absorbs radiation leaving the top and sides of the core, so that power station staff can work safely around the reactor.

- It is convenient to be able to refuel the reactor while it is in use without shutting it down. In many reactors this is done from above with a special machine (called a charge-discharge or refuelling machine) which removes used or spent fuel elements and replaces them with fresh fuel.

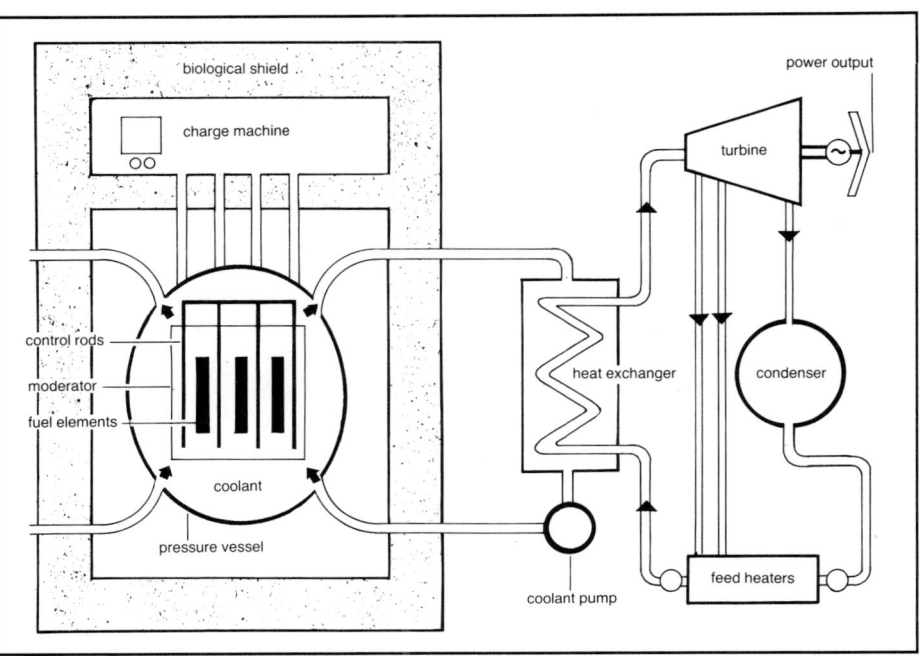

Figure 4, *below: Simplified diagram of a nuclear power station*

Great care has to be taken to avoid any leak of radioactivity from the reactor to the outside world. This radioactivity is mainly from the fission products which build up within fuel elements. Leakage is prevented by using sealed fuel element cans, strong pressure vessels and closed coolant circuits which are carefully tested and checked for any weaknesses. Because fuel element cans could be damaged if they were allowed to overheat, the coolant circuit is very important, and additional emergency cooling systems come into action automatically. At the Three Mile Island power station in the USA, in March 1979, a valve in the coolant piping got stuck in the open position, and the water coolant was lost rapidly. The emergency cooling-systems worked as intended, but confused operators switched them off and, although no one was harmed, the fuel elements were badly damaged.

Different types of reactors   There are many types of reactor and they differ quite a lot in their design. In Britain we have developed reactors using carbon dioxide gas as a coolant with a huge assembly of graphite blocks as the moderator. Fuel elements are loaded into channels in the graphite. These reactors have a large heat capacity, making them relatively slow to heat up or cool down. The first type built were called MAGNOX – named after the metal alloy used in making the fuel cans. Later, stations of the Advanced Gas Cooled Reactor (AGR) type were built.

The recent Sizewell enquiry considered and approved the construction of a Pressurised Water Reactor (PWR), which is already widely used in the United States, Europe and other parts of the world. This has water under pressure, which serves as both coolant and moderator. The Chernobyl reactor used water as a coolant and graphite as a moderator.

How a reactor is controlled   When the reactor is producing a steady power output it is said to be **critical**. How do we increase the power? Simply by raising the control rods. As soon as the power has risen to the desired level, the control rods are returned to their original positions (at which the reactor was critical) to ensure that power remains steady at the new level. Clearly the control rods are a very important part of the reactor, and they need to be able to react quickly.

How can a process as fast as nuclear reactions be controlled safely? The neutrons in the reactor core come from two sources. The majority are prompt neutrons which are emitted at the moment of fission and have lifetimes of less than a thousandth of a second. And a small fraction (less than one per cent) are delayed neutrons from the decay of fission products with time delays of up to 80 seconds. At steady or critical state, the prompt neutrons are topped up by delayed neutrons to maintain a constant rate of fission. The control system therefore depends on these delayed neutrons. And control rod movements within 10 to 20 seconds give adequate control over the chain reaction, so that manual control is quite possible if it proves necessary.

The control rods automatically drop down into the core if an unscheduled rise in power (known as a **transient**) or any other abnormal condition arises. This immediately stops the fission process and shuts down the reactor. The speed of this automatic response shows that if they are properly designed, reactors are stable and straightforward to control.

What went wrong at Chernobyl?

At the time of the Chernobyl accident, the operators had withdrawn the control rods beyond the allowed safety limits and could not shut down the reactor by reinserting them rapidly. This situation arose because the reactor was not being operated at normal power, but at very low power levels for experimental purposes. As fission occurs, neutron-absorbing fission products such as xenon are produced. These products are called neutron poisons. At low power levels these poisons are not destroyed, and they accumulate. The control rods were therefore withdrawn a long way in order to maintain the chain reaction.

Another fault in the design of the Chernobyl reactor was the 'positive void coefficient'. This curious term simply means that as the reactor got hotter the water coolant turned to steam inside the core. Since the steam absorbed fewer neutrons than the water, the number of neutrons increased – yielding yet more energy and heat, and making the

15

problem worse. As a result, the power rose faster than the control rods could be inserted. The effects that led to the Chernobyl accident could not occur in other types of gas-cooled or water-cooled power stations, which are designed so that higher temperatures tend to shut the reactor down.

It can be seen from this brief discussion that the design and operation of a nuclear power station is much influenced by safety considerations. In Britain the safety of reactors is regulated by an independent body, the Nuclear Installations Inspectorate.

Radiation and its effects

Radiation risks can be thought of in much the same way as risks presented by other industrial activities, like those which involve harmful chemicals. Why is it so important to stop man-made radioactive substances (such as fission products) entering our environment, and how do they cause harm? Radioactive substances emit radiation which can damage human body cells. Very large **radiation doses** lead to death quite quickly, or radiation sickness due to damage of the sensitive cells lining the digestive tract. At low doses, the main concern is the possibility of forming cancerous cells.

The International Commission on Radiation Protection has therefore laid down strict guidelines (both for the public and for those working with radiation) to limit exposure to man-made radiation. These limits are enforced in most countries, together with a general principle of keeping radiation doses as far below these limits as can be achieved with reasonable cost. This principle tries to ensure that radiation doses to the public from nuclear power are very small compared with doses from natural or medical uses of radiation. In Britain the National Radiological Protection Board monitors the radiation exposure of the public.

Half-life It is not possible to predict exactly when any radioactive atom will disintegrate or **decay**. However if we imagine a large number of radioactive atoms, the length of time over which half of them will decay is called the **half-life**. After one half-life, half the original atoms will be left. And after a further half-life has elapsed, only half the remaining atoms will be left (i.e. a quarter of the original number). After a further half-life, an eighth will be left, and so on. Half-life is measured in units of time and can vary from seconds for some isotopes to thousands of years for others. The shorter the half-life, the faster the atoms will lose their radioactivity.

Box 2 Different types of radiation

There are four important types of radiation, each with its own characteristics:

alpha-radiation This consists of particles containing two protons and two neutrons in a very stable combination. Because of their mass and double electric charge they can be very damaging to cells. Although stopped by skin, alpha-particles can be harmful if the radioactivity is taken into the body.

beta-radiation This consists of fast (high energy) electrons, and it is not very penetrating. It can give radiation dose to the surface layers of human tissues and within the body (due to radioactivity in food and water or inhalation into the lungs).

gamma-radiation This is like light but of a much higher energy and shorter wavelength, and is invisible to the human eye. It is similar to x-rays, but of higher energy. It is very penetrating, can be stopped by heavy lead shielding or thick concrete, and easily passes through human tissues, leading to radiation dose.

neutrons Fast neutrons are quite penetrating but as they are electrically neutral they do not cause radiation dose directly. They displace protons, and these may then cause cell damage.

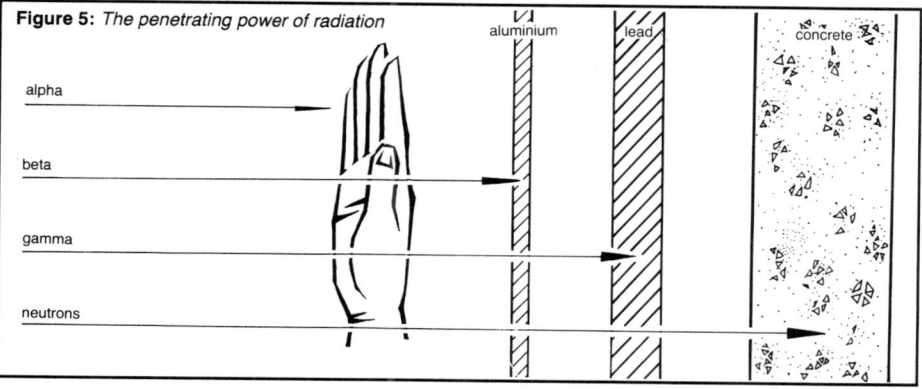

Figure 5: *The penetrating power of radiation*

17

Measuring radioactivity

Radiation arises from disintegrations or decay of atomic nuclei. The unit of radiation is the **becquerel (Bq)** named after Henri Becquerel, a French pioneer in studies of radioactivity. Henri Becquerel discovered radioactivity in 1886 and shared the Nobel prize for physics in 1903. He is also reported to have recognised the biological effects of radiation as a result of burning himself by carrying a vial of radium in his pocket.

Box 3   The becquerel

1 Bq = 1 disintegration per second. The becquerel is a very small unit and hence leads to very large numbers. To avoid this, larger units such as a tera becquerel (TBq) are sometimes used. 1 TBq = 1 000 000 000 000 Bq.

Exposure to large numbers of becquerels does not automatically cause harm. This depends on the kind of radiation and exactly how a person is exposed – the radiation dose. The radiation dose absorbed by the body is measured in **sieverts (Sv)**, named after an eminent Swedish scientist in radiation protection, and includes allowance for some radiation being more effective in causing damage to human tissues. The sievert is a large unit – the average annual exposure of someone living in the United Kingdom is about two thousandths of a sievert, called 2 millisieverts (mSv).

Fortunately, most types of radiation are easy to monitor in the environment. Radiation levels are a small fraction of that due to natural background so any increases caused by man-made radiation may be detected. It is relatively simple to measure radiation because it disturbs atoms. A single particle of radiation passing through a monitor can be detected, for example, in a geiger counter by recording the electric charge produced in gas inside a metal tube.

The nuclear fuel cycle

The fuel cycle refers to the various stages shown in the box – the processing of uranium at the mines, the manufacturing of fuel, power generation, the transport of spent fuel from power stations to Sellafield, and the management of the resulting radioactive wastes.

Box 4 Stages in the nuclear fuel cycle

Mining, milling and extraction
These processes are carried out at or near the uranium mine and result in a pure form of uranium (yellow cake) which is then imported into the United Kingdom.

Fuel production
At special factories in England, uranium metal bars are made and sealed in magnesium/aluminium alloy cans for MAGNOX stations, or the amount of ^{235}U is increased (through enrichment) and uranium oxide is loaded into steel cans for AGR or PWR stations.

Power generation
After the fuel has been used to generate power at power stations, it is discharged into 'cooling' ponds. The fuel stays in the ponds for about a year to allow most of the fission product radioactivity to disappear.

Transport
The spent fuel elements are loaded into specially shielded and protected flasks, which are carried by train to Sellafield.

Reprocessing
The fuel elements are chopped up and dissolved. Unused uranium and the element plutonium are extracted.

Waste management
The most highly radioactive wastes will be sealed into glass blocks. The least radioactive wastes are placed in special trenches. Intermediate-level waste is stored until a means of disposal is ready.

Natural uranium, as it is mined, is slightly radioactive and just under one per cent is ^{235}U. While this proportion is adequate for MAGNOX reactors to operate, AGR and PWR contain more materials in their cores which absorb neutrons, thus decreasing the number of neutrons available for the chain reaction. In these reactors, the uranium must be **enriched** – that is, have the proportion of ^{235}U increased by a factor of about four. This enrichment takes place in special factories where uranium, in the form of a gas, is spun rapidly to separate the lighter ^{235}U from the heavier ^{238}U.

Fuel elements that are carried from the power station to Sellafield for reprocessing are not waste;

they contain useful materials. Unburnt uranium is separated, as is **plutonium**. Plutonium is an artificial element formed in fuel elements from the second isotope of uranium, ^{238}U. The atoms of plutonium (actually ^{239}Pu) can be split to produce power, but plutonium is hazardous to handle and can be used to make bombs, so it must be very carefully safeguarded.

Plutonium can be used as fuel in the so-called **fast-breeder reactor**, and there is a small prototype power station of this kind at Dounreay in the north of Scotland. In such a reactor, roughly the same amount of fresh plutonium is 'bred' (by ^{238}U atoms absorbing a neutron) as is used in the initial fuelling. The overall result is that the total energy obtained from the initial uranium ore is increased by about 50 times. But fast-breeder reactors are rather more expensive to build and will not be practical in economic terms until the price of uranium has risen significantly.

Radioactive wastes from reprocessing (and from power stations) are classified into low, intermediate and highly active wastes. High-activity wastes from reprocessing are not large in volume and can be sealed into glass blocks, which will be stored for about 50 years until they can be disposed of, probably deep below very stable layers of rock. The best ways of storing and disposing of intermediate-level and the larger volumes of solid wastes are now being considered by the nuclear industry. A method which has been used successfully in Sweden, whereby tunnels are excavated beneath stable rocks from the coast out to sea, may be adopted. Low-activity solid wastes (which may be such ordinary items as gloves and other clothing worn during work in the nuclear industry) are larger in volume than the other items of waste. Some are stored; others are disposed of in a special open trench at a site at Drigg in Cumbria. In the future low-activity waste may also be disposed of in coastal tunnels. Storage of all wastes, for example at nuclear power station sites, cannot be the safest long-term approach.

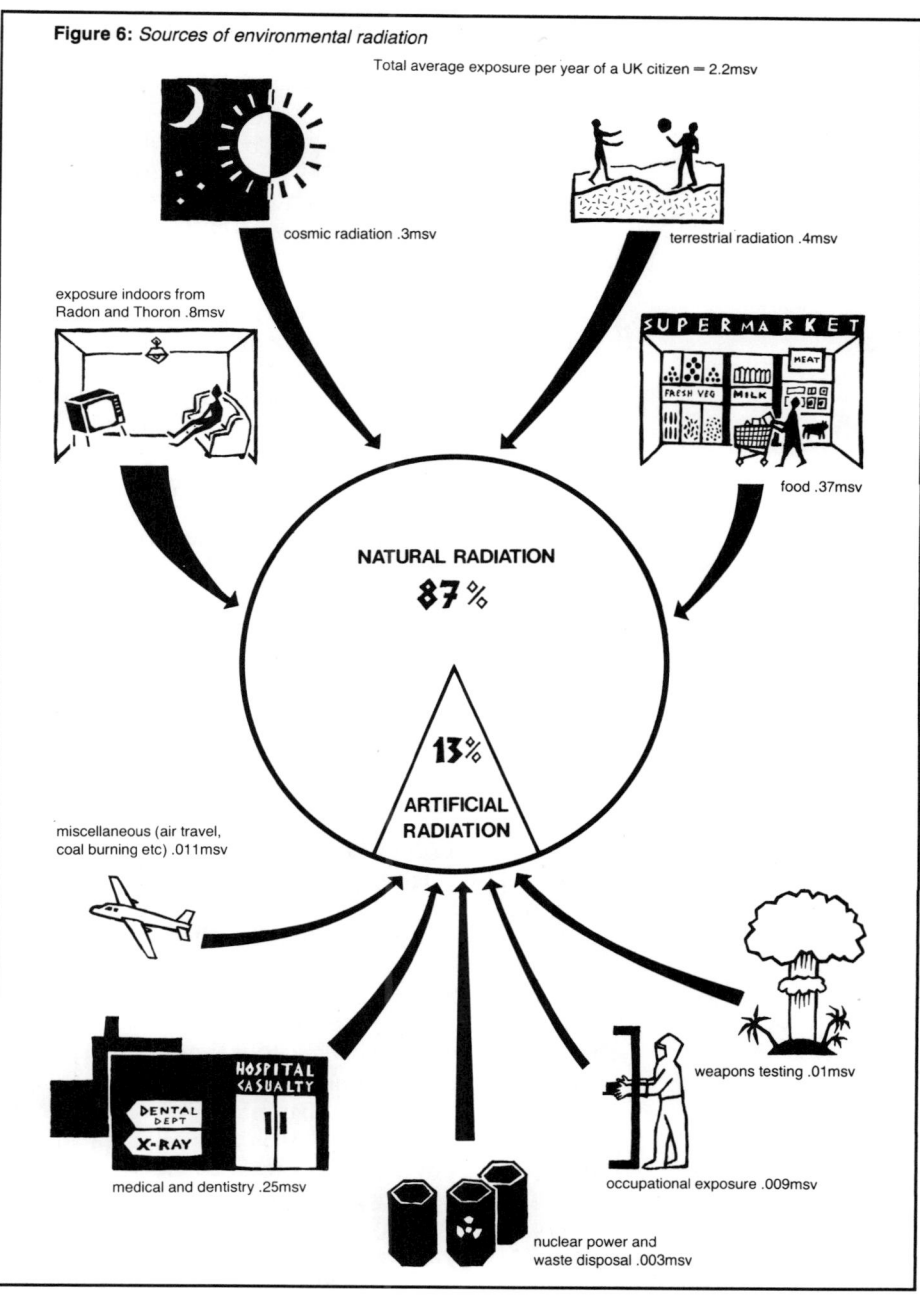

Figure 6: *Sources of environmental radiation*

Environmental exposure to radiation

Everyone is constantly exposed to radiation from natural sources (*see Figure 6*). It can vary quite a lot from one place to another according to the local rocks and soil, which contain different concentrations of naturally occurring radioactive atoms. Most of it comes from **radon gas** escaping into houses from trace quantities of uranium and other elements in the ground. In the United Kingdom about 87 per cent of human exposure to radiation is due to natural sources.

Man-made radiation gives 13 per cent of exposure – mostly from x-rays and medical uses of radioactivity which are justified by the benefits to our health. By comparison, nuclear power causes a very small amount of exposure (about a thousand times smaller), even when allowance is made for radioactive wastes and their treatment and storage.

To put Chernobyl in perspective, the resulting average exposure to radiation in Western Europe over the next 70 years is estimated to be equivalent to a few months of natural radiation.

If nuclear power is to be acceptable it has to contribute minimally to additional exposure. This applies not only to the power stations themselves, but also to radioactive wastes generated and the whole fuel cycle. There must be an overall benefit from nuclear power to counterbalance this additional risk.

Nuclear power in the future

If used with proper care and with wastes properly disposed of, nuclear power is potentially the least polluting major energy source. Unlike the burning of coal, it does not have the side effects of acid rain and possible changes in the climate. However the Chernobyl disaster has reminded us of the importance of high safety standards which must be used to reduce the potential danger to insignificant levels. The future success of nuclear power depends on the public being convinced that safety and radiation protection standards in Britain are very high and that nuclear power offers a real benefit.

Much will also depend on the development and cost of other energy sources, such as coal. If the costs of other energy sources, and the cost of uranium, rise substantially, then fast-breeder reactors should become economic and help us to make much better use of uranium reserves. At present it seems unlikely that they will be widely used before the early decades of the next century.

If fast-breeder reactors are not accepted by the public, there will be a greater need to develop **fusion** power. Fusion is the energy source that powers the sun. In contrast to fission, fusion involves bringing two atoms together (or fusing them). One way of doing this is to confine a high temperature gas of electrically charged atoms (a 'plasma') with the aid of strong magnetic fields. Unfortunately, nuclei repel each other strongly until they are very close together, and so very high temperatures and strong magnetic fields are needed. When two light atoms fuse (for example two hydrogen atoms), the resulting mass is less than that of the two separate particles and so energy is released.

The most promising way of confining a plasma is the so-called TOKAMAK arrangement, where the plasma is in a doughnut-shaped vessel surrounded by sets of conductors, carrying electric current to produce the magnetic field. Although scientists at Culham and other laboratories now seem close to showing that this kind of machine can achieve fusion, it is likely to be the second half of the next century before fusion power becomes practical.

In the immediate future, radioactivity will undoubtedly continue to be an emotive subject, as will nuclear safety and waste disposal. It is only by understanding the benefits and the risks, and through confidence in the safe operation of the nuclear industry, that the trust of the public can be gained.

THE ENERGY 'CRISIS'
Power for our future

Richard Marshall

> **This chapter investigates:**
> - what energy is
> - types of energy
> - how energy is transformed
> - the properties of energy
> - the laws of energy
> - what energy efficiency really means
> - how energy can be stored and saved
> - energy planning for the future

What is energy?

The last chapter discussed nuclear power, one of the most controversial forms of **energy**. Of course there are many different types of energy, but before looking at them, we need to know what energy really is. The first thing to understand is that energy is not a substance, you cannot hold it in your hand. Also, unlike the closely related ideas of force, work and heat, it is not based directly on sensory evidence. Energy tells us about the relationship between one thing and another, and how they might affect each other.

Although energy is not a substance, we can easily recognise its effects. In many cases it is clear that a large quantity of energy must have been involved to produce a particular situation: for example, a large crucible of molten metal, a reservoir full of water high up the side of a valley, a train travelling at high speed or a cylinder of compressed gas. We can also easily observe processes that involve the transformation of energy: for example, a streak of lightning, a tree being blown over in a high wind or a wave breaking upon a beach.

However, sometimes the energy is 'hidden' (or potential), rather than obvious. The most striking example of potential energy is fuel. There are many different types of fuel but they are all naturally occurring sources of easily changeable energy. Other examples of potential energy are a coiled spring (which has sufficient potential energy to power a clockwork toy for several minutes as it unwinds), and the raised weights of a grandfather clock (which can energise the clock for more than a day, as they fall back to their lowest position).

Types of energy By observing the processes in the world around us we can divide energy into types. These are **gravitational, mechanical, thermal, chemical, electro-magnetic** (including **radiant** and **solar** energy) and **nuclear**. Even apparently simple processes often involve transformations of several types of energy.

Energy transformations Some of the **renewable energy** supply systems provide excellent examples of energy transformations in action.

Box 1 Types of energy

Type of energy	Explanation	Example(s)
gravitational	due to differences in height between objects	water in a high reservoir
mechanical	(a) due to stresses and tensions (**elastic** energy)	a compressed spring
	(b) due to the motion of an object (**kinetic** energy)	anything moving
thermal	due to differences in temperature between objects	a tank of hot water
chemical	due to differences in chemical composition between objects	anything that burns
electro-magnetic	due to electric or magnetic forces between objects	attraction or repulsion between two magnets; an electric spark
nuclear	due to differences in mass between two atomic nuclei	radioactive decay; a nuclear bomb

---

Hydro-electric power stations A hydro-electric power station converts some of the gravitational and/or kinetic energy of falling and flowing water into electrical energy. Some hydro-electric power stations operate on a **run of the river system**, which uses the kinetic energy of the

25

moving water to rotate a water **turbine**. The turbine is in turn connected directly to a **dynamo** which generates the electricity. The turbine and dynamo together are known as a **turbo-generator**. Although the river may be dammed to provide a reservoir for times when the flow is low, the height through which the water falls to gain speed is not large, and the system is said to operate with a **low head** of water.

Other hydro-electric power stations operate on a **high head system** which exploits the huge pressure exerted by the water. Most of this pressure is transferred to the turbo-generator blades. Only enough of the gravitational potential energy is converted into kinetic energy to allow the water to flow away.

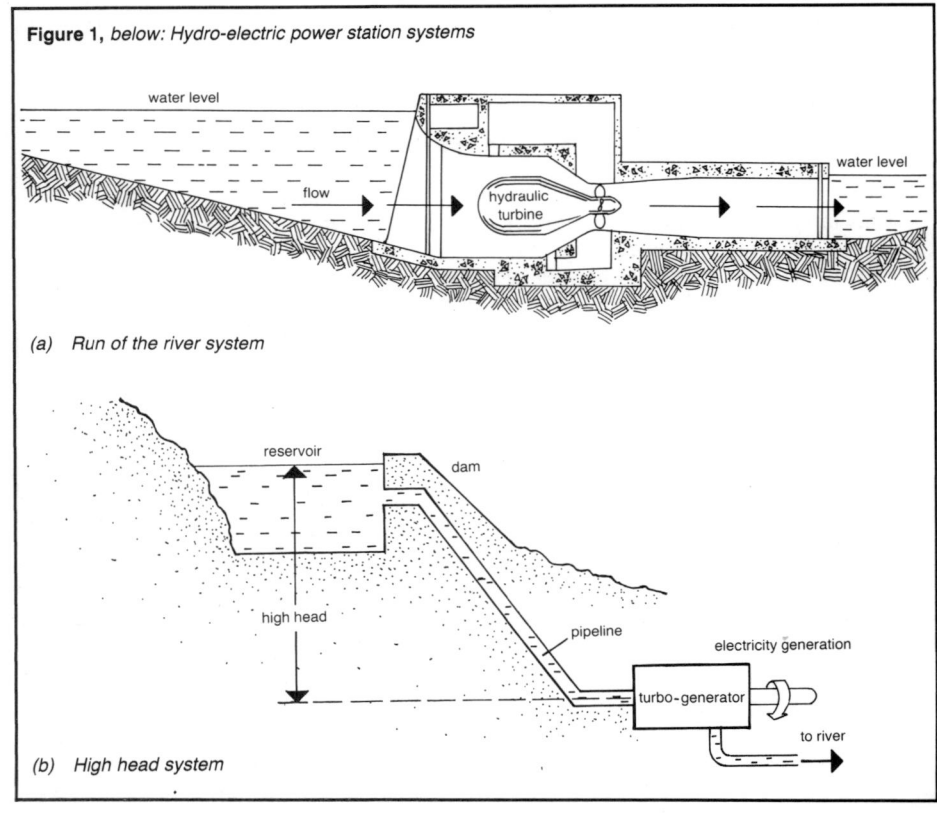

Figure 1, *below: Hydro-electric power station systems*

(a) Run of the river system

(b) High head system

Figure 2, *above: A wind turbine on the Orkneys that will provide electricity for 2000 homes (wind power could provide 20 per cent of the UK electricity demand at a lower projected cost than nuclear power)*

Wind turbines   Wind turbines convert the kinetic energy of flowing air into rotary kinetic energy and then into electricity. The most familiar types look like large twin-bladed aeroplane engines mounted on top of tall towers. Electricity is produced in exactly the same way as in a run of the river hydro-electric system.

Wave power   Winds blowing over open water produce waves. As the waves pass, the water moves up and down, continuously transforming gravitational energy into kinetic energy. One of the most promising wave

Figure 3, *below: Design for an Oscillating Water Column wave power system*

power devices has the waves entering an upturned column-like chamber. Inside the column the kinetic energy of the rising water compresses trapped air which is converted into elastic energy, and this energy is then released through a turbine.

Tidal power Tidal power is at the opposite extreme from wave power. Wave power exploits a fairly small quantity of water moving up and down several times each minute, but tidal power schemes involve huge volumes of water moving up and down approximately twice each day. They convert the gravitational energy into electrical energy using turbines which are similar to the ones found in a run of the river hydro-electric system.

Figure 4, *below: The La Rance tidal power station on the coast of Brittany which has been fully operational since 1967 (the UK has many favourable sites for a tidal power station such as the Severn and Mersey estuaries)*

Energy's remarkable properties

1 *Nothing can happen without in some sense using energy.*

Because energy is involved in *all* situations and processes, it can be measured in many different ways. Calories, therms, joules, ergs, electron-volts, kilowatthours – these are some examples of the units we use to measure energy in different processes.

2 *Different types of energy have a different quality.*

Some forms of energy are more versatile than others. For instance, some processes that are possible using electrical energy are impossible with thermal energy. Different types of energy can be ranked according to their **quality** – that is, the ease with which they can be transformed into another type of energy. This idea of energy quality is closely related to what physicists call **entropy**.

Electrical energy and mechanical energy are examples of the highest-quality forms of energy possible. The quality of thermal energy depends upon its temperature – the hotter the better, if you want to maximise the amount which can be transformed into mechanical energy using an engine. The sun radiates solar energy at a temperature of about 6000 degrees Celsius. Thus, radiant solar energy is also of very high quality and can be efficiently transformed directly into electricity, using a solar cell for example.

It is always possible to convert high-quality energy to lower-quality energy. It is impossible to convert a given quantity of low-quality energy into the *same* quantity of a higher-quality energy. Another way of saying this is that any form of energy can be transformed into the same quantity of thermal energy, *or* a lesser quantity of mechanical energy plus some waste heat.

3 *No matter what happens the quantity of energy remains unaltered.*

When all the different forms of energy that may be involved in a process are taken into account, the total quantity of energy supplied to any device must equal the total quantity leaving it. For example, the input to a hydro-electric power station is the kinetic and gravitational energy of the water that enters the turbine. Electrical energy is the major output, with some kinetic energy in the water leaving the turbine. In addition, the water will be a little warmer because of the heating effect of frictional forces, and there will be some energy radiated as noise from the swirling water and the spinning machinery.

The laws of energy

Our practical knowledge of energy is summarised in two scientific laws, the **first** and **second laws of thermodynamics**. Like all scientific laws, they are valid because, so far, they have not been disproved by experience. They are two of the first principles upon which all other theories are based. Scientists believe that any theory which contradicts either of these two laws must be wrong.

The first law

The first law is a statement of the third property of energy and is also known as the principle of the conservation of energy. This says that 'The total quantity of energy in the universe, although unimaginably large, is nonetheless finite and fixed.' According to this principle, the amount of energy in the universe does not change, and energy cannot therefore be created or destroyed; all that can happen is that energy is moved around. An increase in the quantity of energy in one place must mean a decrease in energy somewhere else. Many increases and decreases take place all the time because energy is always being converted from one form to another. This is partly the sense in which energy is said to be used, and brings us back to the first property of energy: 'Nothing can happen without in some sense using energy.' Another way of putting this is: 'Nothing can happen without energy being changed from one form to another.'

Rather than trying to deal with the whole universe, we can often look at a particular situation in which the boundaries make it impossible for energy to be exchanged. In other words, the total quantity of energy inside the boundaries must remain the same, no matter how complicated the processes occurring within them.

If the boundaries do allow energy transfer to take place, then any change in the energy inside the boundaries must be equal to the total gain or loss passing through the boundaries. The boundaries may be real like the walls, roof and floor of a house, or they may be theoretical, like, for example, an invisible sphere around the planet Earth. The choice of boundary depends on the situation we are interested in analysing.

> **Box 2 Energy inputs and outputs for a house**
>
> *Inputs (+)*
> fuel and power bought by the occupants
>
> body heat from the occupants themselves
>
> solar energy through the windows and warming of the walls
>
> *Outputs (−)*
> thermal energy through the walls, floor, roof, windows and doors
>
> thermal energy in warm water sent down the waste pipes
>
> warm air lost through cracks and gaps, or up the chimney

In the example of a house, the boundaries are obvious and the energy inputs and outputs are easy to identify.

To maintain a steady internal temperature the inputs must balance the outputs. Most of the energy used by the heating system is used to warm the cold fresh air, which comes in through cracks and gaps. Many householders claim that they have very effective heating systems, by which they mean that their houses are kept comfortably warm. But, as we shall see later, domestic heating systems are actually very inefficient in their use of energy.

The second law The second law of thermodynamics is that the quality of energy is always decreasing. In other words all energy transformations involve a loss of energy quality. Unlike its quantity, quality is not conserved, so using energy also means reducing its quality. For example, some of the thermal energy of hot steam can be converted at a power station into electricity, but if the steam is allowed to cool, then virtually none of it can be converted into electricity. The quantity of energy is the same, but nearly all its quality is lost when it cools down.

Engines The second law requires that work output from an engine (something that converts one form of energy into another) must be quite a lot less than the heat input. Perfect, or equal, conversion is impossible. The maximum conversion efficiency in the most modern power station is about 40 per cent. To put it another way, more than 60 per cent of the thermal energy input cannot be converted into the work output.

Natural heat engines and renewable energy supplies

In nature, thermal energy is continually being converted into mechanical energy, and these processes form the basis for many of the renewable energy technologies being developed today. On the simplest level, a sailing ship uses the kinetic energy of the winds. Winds are produced by the unequal heating effects of solar energy on the air, causing the air to move. Wind turbines are another way in which we use this natural heat engine.

The natural conversions are rather more complicated in the processes that produce the rainfall which is eventually used by a hydro-electric power station. First, solar energy warms the water on the Earth's surface, causing it to evaporate and rise into the atmosphere, thus gaining gravitational energy. The water vapour is blown over the land by the winds, and then condenses into rain. The rain uses some of its gravitational energy in falling on to the land above sea level, and collects in lakes and rivers. Hydro-electric power stations then use some of this remaining gravitational energy to produce electrical energy.

Wave power devices use the wave motion generated by winds blowing over open water. Wave power is thus solar power twice removed – solar energy produced the winds and the kinetic energy of the winds produces the waves.

Summary of the first and second laws

The first law of thermodynamics tells us that there is a fixed quantity of energy in the universe, and that thermal energy and mechanical energy are different aspects of the same thing. However, the second law states that the quality of energy decreases in any conversion, and the differences between thermal and mechanical energy are crucial.

Perpetual motion machines

Despite the considerable and continual efforts of countless inventors over the centuries, nobody has yet been able to design a perpetual motion machine; if they did, they would disprove the two laws. Remember that a machine is a device that provides a continuous mechanical or electrical output – that is, an output in the form of the highest-quality energy possible.

There are two quite different theoretical possibilities for a perpetual motion machine. Some designers propose a device that can replace the energy output of the machine from some inexhaustible source inside the machine itself. Such a machine would show perpetual motion of the first kind because its operation would contradict the first law. The second kind of perpetual motion machine would keep going by recycling the initial input of energy in such a way that it could be repeatedly restored to its original quality for re-use, without any other energy input. This would contradict the second law.

If energy is conserved what is the energy 'crisis'?

In the last 15 years or so, there has been much talk of an energy crisis, and pleas from the government to save energy. If energy is always conserved what does this mean? To put it another way: if energy is conserved then all energy conversion processes must be 100 per cent efficient. This puzzle can be answered by analysing what we mean by efficiency. Each of the laws of thermodynamics suggests a different way to define efficiency and we can illustrate this by looking again at our home heating systems.

The efficiency of home heating

A typical household central-heating system has an energy efficiency of 70 per cent, whereas its thermodynamic efficiency is a miserable 5 per cent at best! This shows the enormous room for improvement, particularly when you realise that about 25 per cent of all the energy utilised each year in the United Kingdom is used to keep homes warm.

Why are the two values so different? What we actually want from the heating system is a house warmed through to around 20 degrees Celsius. The energy quality of the thermal energy in a warmed house is quite small compared to the energy quality in the fuel originally supplied to power the central-heating system. This fuel is burnt at a temperature of around 2000 degrees Celsius to provide a heating level which is many times lower! If the high-quality thermal energy released by burning the fuel was used in an engine, there would have still been

> **Box 3 The meaning of efficiency**
>
> Our discussion of energy suggests that there are two ways to define the efficiency of any energy conversion process. The first law tells us that 'what goes in must come out (or stay inside)'. Not all the types of energy output are wanted, and so we can define the energy efficiency as:
>
> $$\text{ENERGY EFFICIENCY} = \frac{\text{QUANTITY OF ENERGY OUTPUT OF THE TYPES YOU WANT}}{\text{TOTAL QUANTITY OF THE ENERGY INPUT}}$$
>
> The energy efficiency considers the quantities of energy, and does not keep track of the loss of energy quality. From the point of view of the second law of thermodynamics, the best process would be one that results in the smallest loss of energy quality. Thus the thermodynamic efficiency is defined as:
>
> $$\text{THERMODYNAMIC EFFICIENCY} = \frac{\text{ENERGY QUALITY ACTUALLY USED}}{\text{THE MINIMUM ENERGY QUALITY NEEDED}}$$

enough waste heat to heat the house.

However, using high-quality fuel is a very convenient way to heat our homes, so perhaps the best way of saving energy quality is to make sure we use as little of it as possible. It is easy to design and build housing that is so well insulated that the warming effects of the occupants themselves, the lighting, television, hot water system, kitchen appliances and the like, provide enough heat for most of the year. Not until the outside temperature falls several degrees below freezing is the actual heating system needed at all.

The meaning of the energy 'crisis'

It is not energy as such which is likely to be scarce, but the chemical fuels that can be easily converted into high-quality mechanical energy, which in turn can be transformed into electrical energy. Our way of life has become very dependent upon adequate supplies of such fuel. Oil and gas, in particular, are expected to become scarce in the foreseeable future. Thus, from a practical standpoint, energy conservation means reducing the rate at which we lose energy quality from traditional energy sources.

Fortunately solar energy is very high-quality energy and is naturally converted into the highest-quality forms in the environment. Furthermore, it is continuously distributed all over the planet at a rate that can easily satisfy our most inflated estimates of future energy demand.

Quite a lot of energy quality is lost through our use of fossil fuels because of the problems caused by variations in our demand for energy. However this variation in demand is not such a problem if the type of energy we want is easy to store.

Storing energy

Fuels are natural stores of chemical energy that can be converted into other forms of energy when they are needed. Thermal energy can be stored by having insulated containers for the substance or environment (e.g. the inside of a house) that we want to keep warm. Unlike thermal and chemical energy, electricity cannot be stored in bulk, so we have to convert it into another form of energy and then back into electricity when it is needed. A car battery stores a small amount of energy chemically. Storing electricity on a large scale involves the use of a special type of hydro-electric power station.

When the demand for electricity is low, the surplus output of power stations is used to pump water up to an elevated reservoir. Later, when demand is high, the reservoir can be used to provide hydro-electric power. The UK already has several of these **pumped storage schemes**. An added bonus of such schemes is that they also save energy although it is a little harder to see why.

Saving energy by storing electricity

The key fact is that the more modern our power stations, the better their energy efficiency. As the demand for electricity rises during the day, the newer, more efficient power stations are used in preference to the older, less efficient ones, so it is the less efficient stations that have to cope with the peak demands. This difference in energy efficiency (i.e. the difference in their fuel input for the same output) is large enough to make it worth investing in pumped storage systems, because the cost of construction and operation is more than paid for by the savings from using the more efficient power stations for longer periods.

When the demand is low, the most energy efficient stations that would otherwise be idle are used to pump water uphill from a lower to a higher reservoir. When demand rises and all the newer power stations are being used, the water stored in the upper reservoir can be released to generate hydro-electric power as it returns to the lower reservoir. A single mechanism acts as both pump and turbo-generator, as required.

If energy is conserved where does it all go?

Eventually all the energy we get from fossil fuel becomes low-quality thermal energy or waste heat. Although it is then of no practical use to us, it can still have far-reaching effects on the way we live. If we dump too much energy too quickly into our environment we will cause changes in the world's climate. So our real problem in the future may well spring from using too much energy (as far as our environment is concerned) rather than not having enough energy (as far as our world economy is concerned). We need to learn that the economy and the environment are not separate concerns.

The problem of too much energy

Why doesn't the average temperature over much of the Earth vary that much from year to year? The Earth's only energy contact with the universe is an input of solar energy, and the **infra-red radiant energy** it in turn radiates into space. For the average temperature to be constant these two energy flows must be equal. If more energy were to arrive on Earth (from *any* source) than the amount radiated away, the average temperature of the planet would have to increase until its infra-red radiant energy output balanced the extra input.

There are two unavoidable side effects of burning fossil fuels. First, virtually all thermal energy released by their combustion warms the surface of the Earth in some way. Secondly, and much more importantly in the short term, the burning of fossil fuels has an effect on the atmosphere. The main constituent of fossil fuels is carbon, and burning carbon produces the invisible gas, carbon dioxide.

Unfortunately carbon dioxide is fairly **transparent** to the incoming solar radiation but rather **opaque** to the outgoing infra-red radiant energy from the Earth. This means that the input of solar energy becomes greater than the output of radiant energy, and upsets the balance of energy flows which determine the average temperature of the Earth. This is popularly known as **the greenhouse effect**. (Real greenhouses usually keep warm in a rather different way – but that's another story!)

The planet responds by warming up until its increased rate of infra-red radiation can establish a new energy balance. The detailed argument is more complicated than this. The temperature changes will not occur evenly over the Earth's surface and there are other gases such as **fluorocarbons** from aerosols which add to the effect. The details of what will happen to the global climate are essentially unpredictable and are only reversible over centuries (if at all).

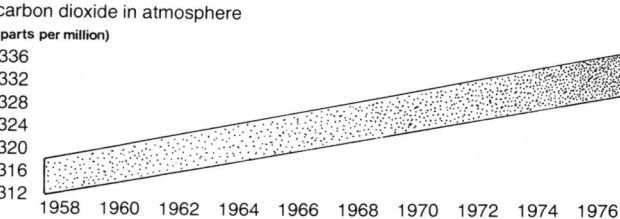

Figure 5, right: Carbon dioxide concentration in the atmosphere

How can we avoid the energy 'crisis'?

First we must make sure that we use all fuels as efficiently as possible. This will minimise the side effects they are bound to produce, and also make the task of meeting our future demand for high-quality energy that much easier, no matter which sources we use.

Secondly, we can avoid adding to the problem by making more use of **renewable energy** supplies such as solar power in all its forms (its direct warming effects and indirectly as wind, wave and hydropower). Our global environment is already adapted to these, so exploiting them does not add an extra energy flow.

Nobody disputes that, in the end, renewable energy sources are going to have to account for a large proportion of our energy needs. The debate only concerns how soon this must happen. The global environmental problems being caused by traditional fuels suggest the sooner the better.

37

3 TEST TUBE BABIES
Creating new life

Virginia Bolton, John Parsons and Stuart Campbell

> **This chapter investigates:**
> - normal reproduction
> - causes of infertility
> - *in vitro* fertilisation
> - new technology
> - future possibilities

The world's first test tube baby, Louise Brown, was born on 25 July 1978. Her birth was the culmination of ten years of painstaking research by two men who are now household names, Robert Edwards and Patrick Steptoe.

Test tube babies are produced by a process called ***in vitro* fertilisation**, which means literally 'fertilisation in glass', as opposed to *in vivo*, which means 'in the body'. In practice, eggs are removed from the woman's body and placed in tubes or dishes (usually plastic, not glass!), containing a special liquid (culture medium). The sperm are added to the eggs in these tubes or dishes, where fertilisation takes place, and the resulting embryo(s) are then put into the woman's womb (or uterus).

This chapter explains *in vitro* fertilisation (IVF), and examines the implications of this, and other types of reproductive technology, for the present and the future.

Normal reproduction

Sperm and eggs (collectively known as **gametes**) transfer the genetic characteristics of the man and woman to their child. It is our **genes**, strung together in chains called **chromosomes**, which make us a unique individual; they programme the development of the baby in the uterus, determining features like eye and hair colour.

All human cells (except sperm and eggs) contain 46 chromosomes arranged in 23 pairs. One of these pairs, the sex chromosomes, consists of either two X chromosomes (producing a female), or an X and a Y chromosome (producing a male). When the gamete is produced, the chromosome pairs split, so that each sperm and egg carries 23 chromosomes (half the normal quota of 46). This means that although all eggs carry an X chromosome, half the sperm will carry an X (and therefore result in a female baby) and half will carry a Y (thus resulting in a male). Moreover, before the chromosome pairs split, they exchange genes randomly, making each gamete genetically unique.

Sperm production and transport

Large numbers of sperm are produced continuously in minute tubes inside the testicles, from puberty until death. They take 60 days to develop, and are at first unable to move. Later, they become **motile** (moving rather like a tadpole) in the **epididymis**, a coiled structure next to the testicle.

During orgasm, the highly concentrated sperm solution is diluted with **seminal fluid**, and is then forced along the **urethra**, out of the man's penis and into the woman's vagina.

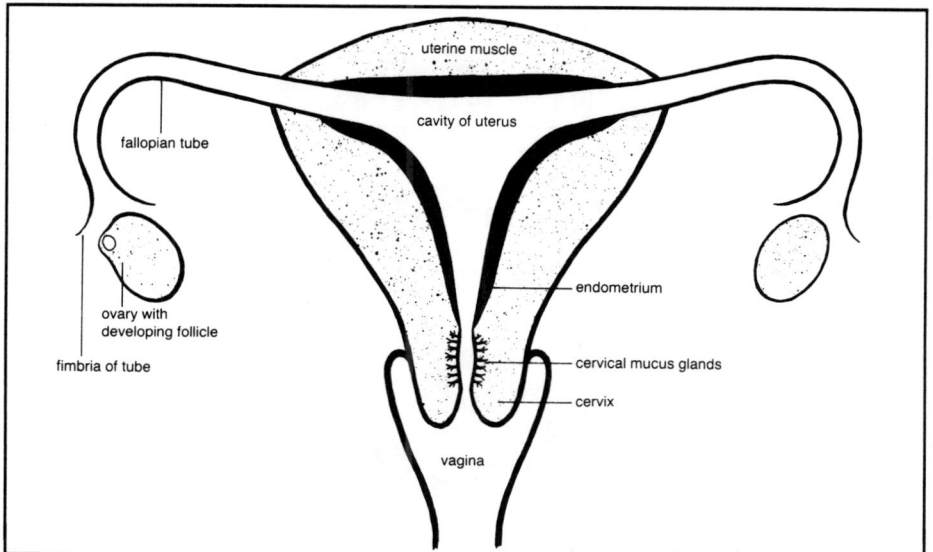

Figure 1, below: Simplified diagram of the female reproductive organs

Inside the woman, the liquid semen thickens, and then after about 30 minutes it returns to a runny consistency. The glands at the **cervix** (or entrance to the womb) produce mucus. During most of the menstrual cycle this mucus is impenetrable, but when an egg is about to be produced, it becomes clear and runny, like egg white, and the sperm can swim through it easily. Aided by muscular contractions, the sperm pass rapidly through the uterine cavity, into the **fallopian tubes**.

Egg production and transport

Egg production is very different from that of sperm. All women are born with a limited supply of eggs in their ovaries. These eggs are constantly being released in small numbers, but if the required **hormones** are not present, they will die. A woman's reserve of eggs therefore steadily decreases throughout her life.

Each egg is surrounded in the ovary by specialised cells which form the **follicle**. The pituitary gland (the master gland underneath the brain) produces follicle-stimulating hormone, and the presence of this hormone causes the follicle to grow. The follicle cells multiply and in turn produce another hormone, oestrogen. This oestrogen passes into the blood vessels and has effects on the whole body. It is this hormone that causes the lining of the uterus to thicken, and the mucus to become clear and runny.

When the follicle measures 15 to 25 millimetres in

Figure 2: *The follicle*

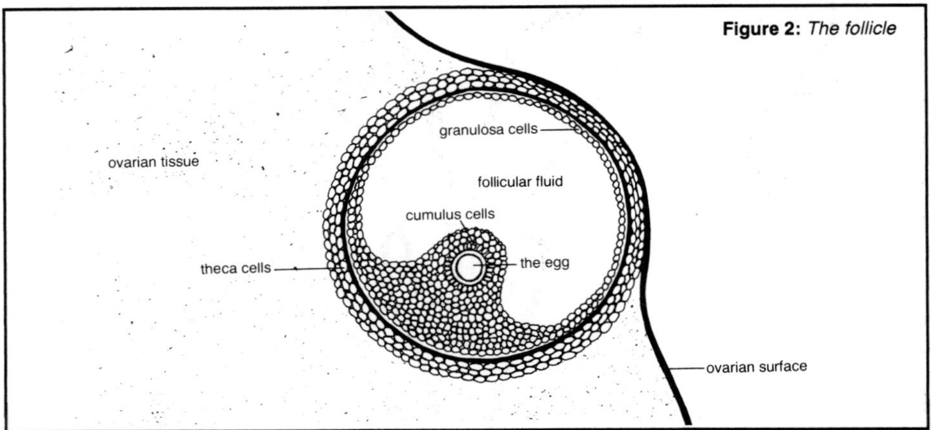

ovarian tissue

granulosa cells

follicular fluid

cumulus cells

theca cells

the egg

ovarian surface

diameter, the pituitary gland releases luteinising hormone. In response to this hormone, the egg finally matures and is **ovulated** (or released from the follicle) 34 to 36 hours later.

The ovulated egg is surrounded by a mass of much smaller 'nursing' cells known as **cumulus cells**. It is picked up by the end of the fallopian tube where fertilisation can take place.

Fertilisation and implantation

During their journey from the vagina, the sperm release **enzymes** which digest away the cumulus cells so that the sperm can reach the egg. Normally, as soon as a single sperm penetrates the egg, a reaction stops further sperm from entering.

The fertilised egg, now called the **pre-embryo**, carries the correct number (23 pairs) of chromosomes. During the following 12 to 20 hours, the chromosomes from the sperm and the egg each form a **pronucleus** which then join together. Thus, the pre-embryo contains genes from both the mother and father in a unique combination, and the baby's individuality is established.

The pre-embryo begins to divide, and 24 to 36 hours after fertilisation it will have reached the two-cell stage. At the two-, four- and eight-cell stages the cells are only loosely attached to each other, but when there are about 16 to 32 cells they become more tightly packed. A little later (about five days after fertilisation), the centre of this ball of cells becomes filled with fluid.

The outer layer of cells will become the **placenta**. Only a few cells, termed the 'inner cell mass', will develop into the fetus. During this period of early development, the pre-embryo will have travelled along the fallopian tube, to reach the uterine cavity. Before implantation can occur, the pre-embryo must 'hatch' from the egg's outer protein covering.

At the same time, in the ovary, the ruptured follicle is producing another hormone, progesterone, as well as oestrogen. Progesterone's function is to prepare the uterine lining for the arrival of the pre-embryo. When the pre-embryo arrives it 'burrows' into the lining of the uterus.

The pre-embryo now releases a hormone which stimulates further production of progesterone by the

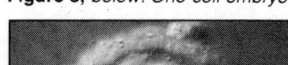

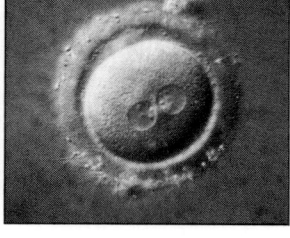

Figure 3, *below: One-cell embryo*

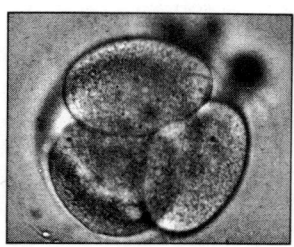

Figure 4, *above: Four-cell embryo*

ruptured follicle, thus maintaining the uterine lining. If there is no pre-embryo or if it is defective the ovary does not receive this hormone, the follicle stops functioning after 10 to 12 days, and the uterine lining is lost in the menstrual flow.

The pre-embryo is not fully implanted until about 14 days after fertilisation. And it is only at this point, when the cells destined to become the fetus begin to develop, that it is called an **embryo**.

Causes of infertility

Eighty per cent of couples will succeed within six months of trying to conceive, and 90 per cent will succeed within a year. For this reason, infertility is not usually investigated until the couple have tried for two years. If the woman is over 35 years old, or if relevant problems are suggested by their medical history investigations may be started earlier.

Although most couples are concerned about controlling their fertility, for about one in five couples the reverse is true. Problems in production and transport of sperm and eggs, or difficulties in fertilisation and implantation can reduce fertility or lead to absolute infertility, depending on the severity of the problem. For instance, the gametes could be abnormal in some way or the pre-embryo's progress could be blocked. If one partner has a minor problem the other partner may be able to compensate, but if both have a problem this might not be the case. Thus, it is not uncommon to find that both the man and woman contribute to their infertility.

Damaged fallopian tubes

Some women are born with abnormal fallopian tubes, but damage to the tubes is usually caused by sexually transmitted diseases or surgery. Damage can range from sub-microscopic destruction of the tiny hairs used in transporting eggs to blockage because of scar tissue.

In about 15 per cent of couples complaining of infertility, the problem is due to damaged tubes. Until recently, surgery was the only hope for such patients, and many underwent operations which had little chance of success. *In vitro* fertilisation was developed to help such people.

Sperm disorders   Sperm abnormalities are found in 25 per cent of couples complaining of infertility. When a semen sample is assessed for its 'normality' the three most important considerations are: the number of sperm present, the percentage which are **motile**, and the number which have a normal shape. The World Health Organisation has defined a normal sample as one in which there are at least 20 million sperm per millilitre of semen, and at least 40 per cent are motile, and of which at least 40 per cent have the classic tadpole shape.

Unfortunately, measuring number, movement and shape of sperm only identifies a possible fertility problem. No test has yet been developed that can distinguish accurately between a fertile and an infertile semen sample. Many men who, according to the measurements described above, have semen abnormalities nevertheless father children naturally. However, many do not, and IVF can help in some cases.

The idea behind IVF helping couples with a semen abnormality is that the sperm are placed very near the egg, and therefore do not have to make the journey from the vagina to the fallopian tube. However, the problem is not always so straightforward. The diagnosed abnormality of number, movement or shape may mean that those sperm are in fact incapable of fertilising eggs, even when they are 'assisted' by mixing eggs and sperm. If this is the case, then current IVF techniques cannot help.

Endometriosis   In some women the lining of the womb grows in the wrong place, and we do not fully understand why this happens. This condition is known as **endometriosis** and it can be treated with drugs, surgery, or both. If these fail, IVF may be used and, particularly in the less severe cases, can give good results.

Ovulation failure   In more than 20 per cent of couples complaining of infertility the woman is found not to be ovulating or to be ovulating rarely. Such women generally do not need IVF and are treated very successfully with drugs which induce ovulation.

Unexplained Infertility — In about 28 per cent of infertile couples, routine investigations do not provide a clear explanation of the problem. Further investigations may reveal subtle abnormalities, but the infertility usually remains unexplained or poorly explained. These patients may benefit from IVF and their chances of success are at least as good as patients being treated because of damaged tubes. IVF also gives useful information about sperm and egg function, and in a significant percentage of unexplained infertility cases the man is found to have a previously undetected problem with his sperm. Couples are then able to consider donor insemination or accept childlessness, rather than go on trying to conceive naturally, unaware of the low chance of conception.

In vitro fertilisation

Ovarian hyperstimulation and monitoring — In IVF the chances of success are greater if several pre-embryos are transferred into the womb, rather than just one. The transfer of one pre-embryo leads to pregnancy in less than 10 per cent of cases but with three or four pre-embryos the pregnancy rate exceeds 30 per cent. To obtain more than one egg, IVF patients are given drugs to stimulate the development of more follicles.

Egg collection — The mature eggs must be collected from the follicles just before ovulation. To make the timing of egg collection easier, most IVF clinics give another hormone which mimics the effects of luteinising hormone. The eggs are then collected 34 to 38 hours later, just before spontaneous ovulation.

Eggs are collected from the ovary by puncturing the follicle with a needle, and sucking out the contents into a test tube. The needle may be guided either by direct vision during **laparoscopy**, using a special telescope, or indirectly, using ultrasound. Laparoscopy was used at first but it has disadvantages in that the patient needs a general anaesthetic, and many patients' ovaries are inaccessible because of scar tissue.

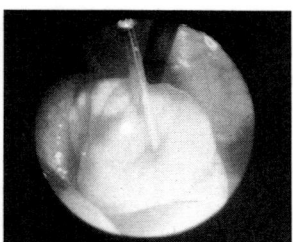

Figure 5, *above: Needle being introduced into the follicle*

On the other hand, **ultrasound directed follicle aspiration** does not require a general anaesthetic and can reach all women's ovaries.

On average, 5 to 7 eggs will be collected from each patient. Occasionally, particularly in older patients with few follicles and a poor level of oestrogen, no eggs will be found. But in some patients who have responded dramatically to the superovulation drugs, over 20 eggs are collected.

Sperm preparation The man's semen is produced by masturbation into a sterile container and allowed to liquefy. The seminal fluid may contain bacteria and debris, so the sample is first washed in a culture medium containing antibiotics, and the motile sperm extracted.

A number of techniques have been developed for separating motile sperm from the rest. The simplest of these is 'wash and swim-up', which involves diluting the seminal fluid with culture medium centrifugation. Fresh culture medium is then placed on top of the solution and the mixture is **incubated** at body temperature for about an hour to allow the motile sperm to swim into the upper layer.

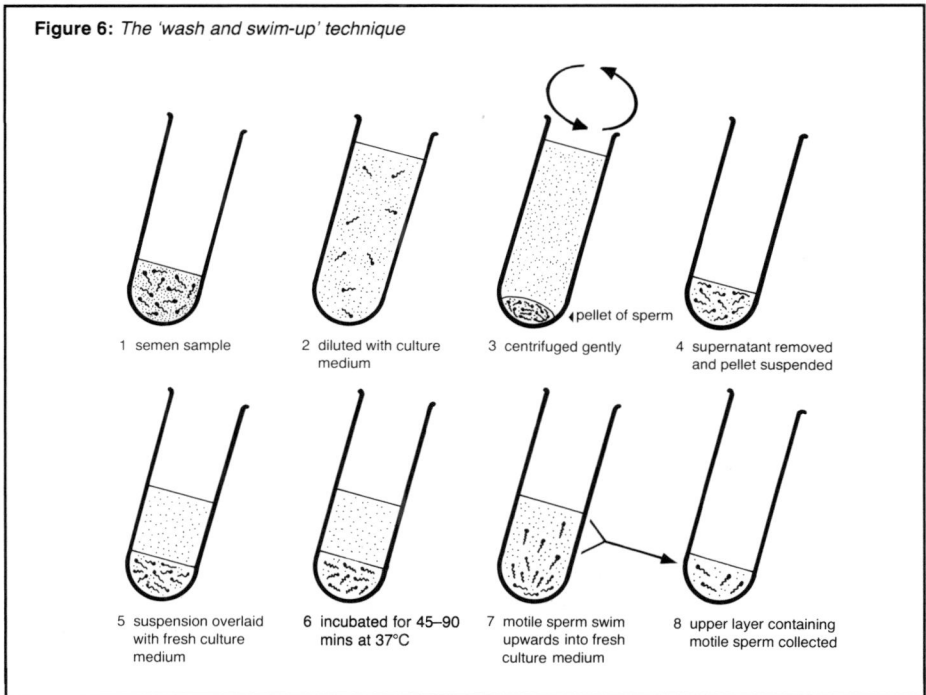

Figure 6: *The 'wash and swim-up' technique*

1. semen sample
2. diluted with culture medium
3. centrifuged gently
4. supernatant removed and pellet suspended
5. suspension overlaid with fresh culture medium
6. incubated for 45–90 mins at 37°C
7. motile sperm swim upwards into fresh culture medium
8. upper layer containing motile sperm collected

With a normal semen sample, this technique will produce sperm that are between 90 and 95 per cent motile. However, it will not separate normal from abnormal sperm, nor will it yield a large number of sperm where the motility of the original sample is low. With very abnormal samples, the 'swim-up' yield will often be so low that sperm have to be salvaged from the lower layer. And a poor 'swim-up' yield frequently indicates that *in vitro* fertilisation may not be successful.

Among other techniques for sperm preparation is 'swim down', which involves layering semen over glass wool, and allowing the sperm to swim through this into culture medium below. Some people claim that this produces a preparation particularly rich in sperm that carry the Y chromosome, and will therefore produce males. Theoretically, since the X and Y chromosomes differ greatly in size, it should be possible to develop such a technique.

The process of in vitro fertilisation

Having been collected, the eggs are incubated at body temperature in individual drops of culture medium. This medium contains very precise concentrations of salts and nutrients, and has to be maintained at the correct level of acidity.

Eggs are collected from the follicles about two to three hours before the time at which ovulation would occur naturally. They are therefore left in the incubator for several hours in order to undergo the final stages of maturation before being exposed to sperm.

During **insemination** 50 000 to 100 000 motile sperm are introduced into each drop of culture medium containing an egg, and after 16 to 18 hours the eggs are examined for signs of fertilisation.

Each egg is freed of surrounding cumulus cells, by being gently sucked up and down a fine glass pipette. We know that fertilisation has occurred when two pronuclei can be seen within the egg. The pre-embryos are then placed in fresh drops of culture medium, where they can be left to undergo the first few cycles of cell division before being transferred to the uterus.

Pre-embryo transfer   About two days after egg collection, the pre-embryos will consist of two or four cells. Up to four pre-embryos are collected simultaneously into a fine, soft, flexible tube which is then passed carefully through the cervix and into the uterus. The procedure is painless and takes less than five minutes, although the patient usually rests for a short while before resuming her normal activities.

Even in the best hands, this method is only about 30 per cent effective. In an attempt to improve the chances of implantation, drugs may be given to stimulate the ruptured follicle to produce more progesterone, or alternatively, the patient may be given progesterone itself. However there is no definite evidence that these drugs help.

Embryo freezing   After the embryo has been transferred to the uterus, the couple may wish to have any extra pre-embryos frozen and stored for possible transfer in a later menstrual cycle. This involves placing the pre-embryos in a protective medium before cooling. The frozen pre-embryos can be stored for many years.

The freezing technique increases the chances of a couple achieving pregnancy from a single egg collection procedure, and has the advantage that pre-embryos are transferred into the uterus during a normal menstrual cycle, without hormonal stimulation. However, it has not yet been perfected. For example, many pre-embryos will not survive the freezing and thawing procedure, and although freezing has been performed at various stages of development, we still do not know which has the best survival rate.

Success rates   The true success rate of IVF is measured as the 'take home baby rate' – in other words, the percentage of couples that have a healthy baby delivered. However the figures are often misleading, since some patients do not complete the treatment or end up losing the pregnancy for various reasons. When advising couples who are considering IVF treatment, the important factors are age, and cause of infertility. Young women with damaged tubes or unexplained infertility, with partners who have normal sperm, have the best chance of success.

New technology

Egg collection and *in vitro* fertilisation have introduced a number of new possibilities for the treatment of infertility, some of which are discussed here.

Pre-embryo research

IVF frequently produces 'spare' pre-embryos. If these are not frozen or disposed of, they can be used in research, but this also leads to heated debate on the ethical issues.

Box 1 The ethics of pre-embryo research

Should a pre-embryo have the same rights as a living person?

Many people believe that human life is sacred at all stages of development, and others that allowing experiments on pre-embryos would be the first step on a slippery slope, eventually leading to experiments on the fetus. However it could be argued that the pre-embryo is only the size of a pin-prick, and consists of a collection of cells, most of which will become the placenta. Also, up to 75 per cent of natural conceptions miscarry spontaneously, often without a fetus ever forming, and 90 per cent of pre-embryos generated by IVF will not successfully implant when transferred to the uterus.

Research using human pre-embryos has several potential benefits. First, it may lead to an increase in the success rate of IVF by developing better culture conditions and identifying the 'best' pre-embryos. This kind of research also helps to improve pre-embryo freezing techniques as well as treatment of male infertility. The other major reason for this research is that it helps to extend technology in other areas, such as development of new contraceptives, prevention of miscarriage, and investigation of the causes of placental and other cancers. It can also help us to discover more about diagnosis of genetic disease before implantation in the uterus, and the use of embryonic cells to repair damaged tissue in adults.

Egg donation Women who do not produce their own eggs require a donor if they are to have a baby. Such women may have reached the menopause before the usual time, had their ovaries surgically removed because of disease or may be carrying an inherited disease.

The concept of egg donation has only arisen within the last decade, as a direct consequence of IVF. While sperm donation is relatively easy for both donor and recipient, egg donation not only involves an operation to obtain the eggs, but also requires much of the technology of *in vitro* fertilisation. This is because the eggs will have to be either transferred to the recipient's fallopian tube, or fertilised *in vitro*, and the resulting pre-embryos transferred to the patient's uterus.

Box 2 The ethics of donation

Should parents be legally obliged to tell the child that he or she is a donor child?

Some people believe that the child should be told, otherwise the relationship would be based on deceit, and would therefore lead to emotional stress on both sides. However other people feel that such decisions are a matter for the parents and that telling the child could disrupt the relationship.

Should a donor child have access to information about the identity of his or her genetic parents?

The argument in favour of access to this information is that everyone has a right to know their genetic origins. However it might be impossible to find enough people to become donors if their anonymity were not guaranteed.

Should a couple be able to choose a close relative as donor?

On the one hand, many couples would feel more secure if they knew the genetic and social background of the donor. But on the other hand, having a close relative as the genetic parent could result in confused and tense family relationships.

Should the suitability of potential parents be assessed before pre-embryo donation?

In the usual adoption procedure, potential parents are assessed before approval is given. However there is no assessment for couples who achieve unassisted (or natural) conception.

INSIDE SCIENCE

Eggs can be donated from a number of sources. For example, patients being treated with IVF may have more eggs than they need and be prepared to give some of them away. Alternatively, fertile women who have completed their families may be prepared to donate eggs at the time of sterilisation.

Pre-embryo donation Pre-embryo donation is a kind of 'prenatal adoption' which is suitable for couples who produce neither eggs nor sperm. The pre-embryos usually come from patients who have completed IVF treatment and have some 'spare'. The donation of eggs, sperm and pre-embryos raises numerous ethical questions, some of which are discussed on page 49.

Future possibilities

Egg freezing Freezing eggs is much less successful than freezing pre-embryos. This is mainly because the egg is a single, large cell, whereas pre-embryos contain a number of smaller cells. Water, which makes up about 90 per cent of the volume of any cell, expands on freezing to form ice crystals, which can cause damage. Large cells are more likely to be damaged during freezing.

If the technique of freezing eggs is developed successfully, patients won't have to undergo so many egg collection operations. A large number of eggs could be collected at one time, three or four of them could be inseminated immediately, and the remainder frozen for later use. The freezing of eggs raises fewer ethical questions than that of pre-embryos. With pre-embryos, there are the problems of how long they should be stored, what happens if either or both the parents die, if the parents disagree or cannot be traced, and whether or not a frozen pre-embryo can inherit its dead parents' wealth.

Sperm injection One way to overcome the problem of sperm that fail to fertilise may be to inject them into the egg. This is being attempted in some research centres, although the work is only in the early stages as yet. It is possible to introduce a single sperm into the area between the outer protein covering and the egg membrane, and sometimes the sperm will then fuse with the egg. However, further research is needed before such 'pre-embryos' can be transferred to the uterus, to establish that they are indeed normal.

Diagnosis of inherited disease

It may be possible to ensure that people carrying an inherited genetic disease, such as cystic fibrosis, do not produce a child carrying the disease. This would involve examining some of the cells of pre-embryos generated by IVF, and only replacing those that do not carry genes for the disorder. Apart from preventing the birth of an affected child, this would eventually lessen the incidence of some genetic diseases.

The use of embryonic tissue in treatment of adults

The cells that make up organs in adults only have a very limited ability to repair damage from disease or aging. In contrast, cells at the early stages of fetal development have a remarkable capacity for healing. It may therefore be possible to use embryonic cells, that have just begun to develop into individual organs, to repair the corresponding organ in an adult. For example, adult blood disease may be cured by injecting embryonic blood cells. However this idea remains purely theoretical at present.

Conclusion

Although IVF was originally developed to help women with blocked fallopian tubes, IVF techniques can also be used to treat other types of infertility. Furthermore, the growing of human pre-embryos *in vitro* could lead to a huge advance in our understanding of, and ability to control, both normal and abnormal human reproduction. Nevertheless this new technology has raised serious ethical problems which need to be resolved before we can make full use of this knowledge.

POLLUTION
The big clean-up

Nigel Graham

> **This chapter investigates:**
> - acid rain and its effects
> - pollution of rivers
> - pollution of beaches
> - pollution of tap water

What is pollution?

Most people today are aware that our environment is polluted and whether through television programmes or magazine articles, the issue of pollution is discussed with increasing concern. We have come to realise that pollution can affect us directly – it can affect what we eat, what we drink, the air we breathe and other aspects of our lives.

Pollution takes many different forms, such as air pollution, noise and contamination of tap water. It can be incidental, meaning that it happens occasionally – usually the result of an accident, as in the tragic leak of poisonous gas at Bhopal in India. Accidents such as these are widely reported in the press, but there is another kind of pollution which goes on all the time and attracts much less attention. This is persistent pollution – for instance, the pollution caused by chimney stacks pumping out thick black smoke every day, or the damage caused to the atmospheric ozone layer by the use of aerosol sprays.

This chapter looks at four well-known aspects of persistent pollution, all connected in some way with water. It explains each type of pollution, how it is caused and what can be done to prevent it – or reduce its effects.

Acid rain

What are acid lakes?

Imagine a clear blue lake in Scandinavia. Although it looks beautiful in the sunlight this lake has no snails or molluscs, no salmon or trout, no perch or pike. It may have certain kinds of plankton and white moss – but that's all. This is only one of the many freshwater lakes in Scandinavia and North America which are known to suffer from **acidification**.

> **Box 1 Acidity**
>
> The **acidity** of water – how much acid is in the water – is determined by the number of **hydrogen ions** it contains. A hydrogen ion is a hydrogen atom which has lost its negatively charged electron. This means that it becomes positively charged.
>   The acidity of the water is measured using the **pH scale**, which ranges from 0 to 14. On this scale:
>   0–7 indicates acid water
>   7 indicates neutral water
>   7–14 indicates **alkaline** water

Going back to our lake in Scandinavia – any increase in acidity in a healthy lake can damage animal and plant life. In general, if the pH value drops below 6.5 aquatic life begins to suffer, and all normal life (snails, fish, sensitive insects, etc) is destroyed when the pH is less than 5. If the pH of the lake is between 5.5 and 4.5, it is **endangered**, and below 4.5 it is described as **critically acidic.**

Figure 1, below: Sensitivity of aquatic organisms to a lowered pH in fresh water

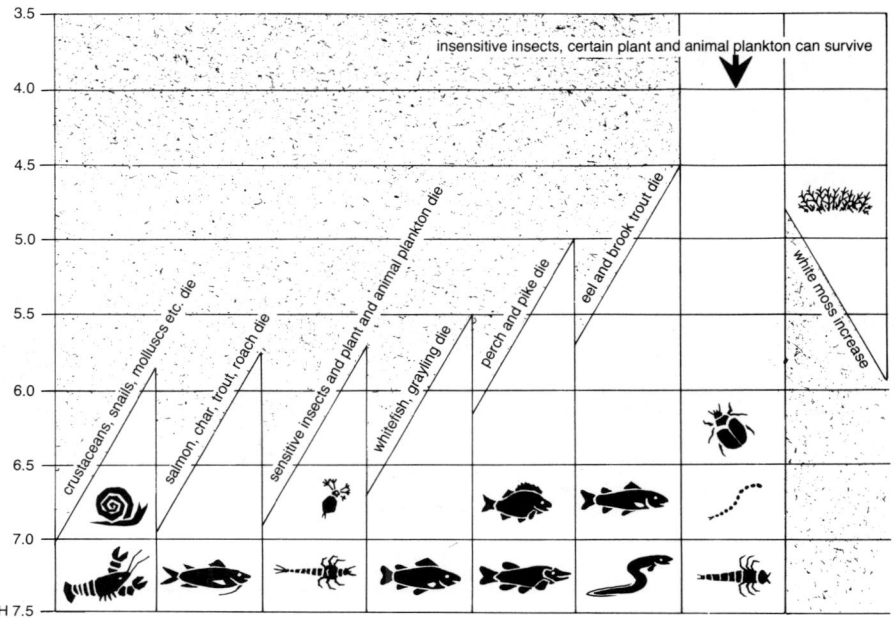

INSIDE SCIENCE

Acidity damages aquatic life in many different ways, particularly by increasing the amount of dissolved metal in the water. This is because certain metals, which are naturally present in the lake, react with the hydrogen ions to produce metal ions. When they are converted into ion form, these metals (especially aluminium) become very poisonous to many forms of aquatic life. For instance, it is probably aluminium poisoning which has caused the mass death of fish in acid lakes. Other dissolved metal ions, such as cadmium, zinc and lead, can also reach harmful levels because of acidification.

What causes acid rain? Some power stations, brick and cement factories, and car engines all work by burning **fossil fuels**, such as coal and oil. When these fuels are used in power stations, factories, engines and other combustion processes, they produce large amounts of sulphur and nitrogen gases which are released into the air. Eventually some of these chemicals form **acid rain** in the atmosphere.

What is acid rain? The sulphur and nitrogen gases combine with oxygen and water vapour in the atmosphere to form

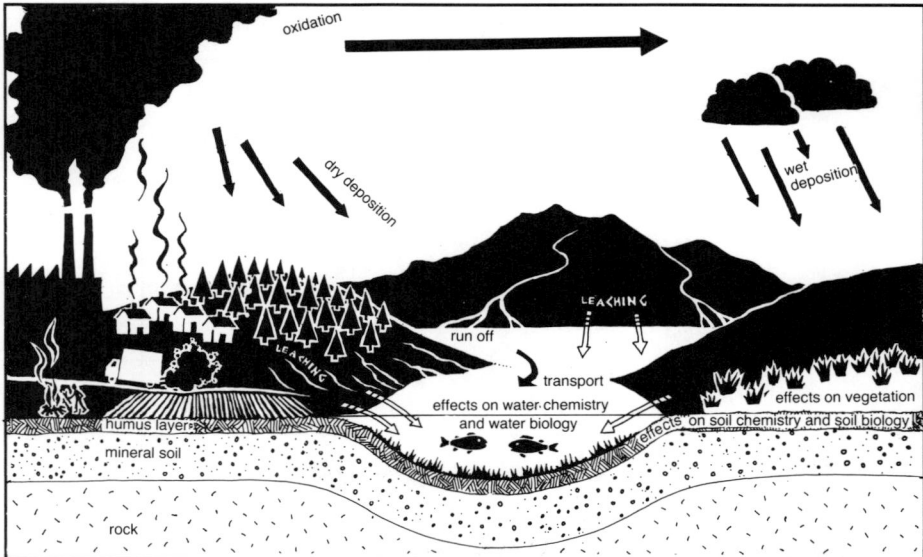

Figure 2: *Environmental effects of acid rain*

sulphuric acid and nitric acid, which then dissolve in rain droplets. Natural rainfall has a pH of about 5.65, so acid rain is rainfall with a pH of less than 5.65.

When the sulphuric acid and nitric acid reach the soil, they combine with dissolved metals (especially aluminium) and damage the roots through which trees absorb much of their nutrition. To make matters worse, important nutrients, such as calcium, magnesium and potassium, are dissolved by the acidic water and washed out of the soil. The result is that leaves shrivel up and large areas of forest begin to die. Experts believe that the current damage to the Black Forest in Germany is largely due to the nitrous and sulphurous gases produced by European chimneys and car exhausts.

How can the environment resist acid rain? The environment does have some natural defences against acid rain. Both soil and water contain **buffering ions** or salts, such as bicarbonates, borates, silicates and phosphates. These alkaline buffering ions **neutralise** the acid in the rain, thus preventing acidification.

How can we help? Some lakes don't have enough buffering ions to resist acidification, and in these cases, alkaline substances can be added to neutralise the acidity. Lime (or calcium oxide) is often used in this way because it dissolves in water to form an alkali made up of calcium ions and hydroxide ions. The hydroxide ions neutralise the acidity in the water, and this process is known as liming.

Lime also reduces the amount of dissolved metal in the water. For example, the positively charged aluminium ions attract the negatively charged hydroxide ions to produce aluminium hydroxide which is a solid substance. (This type of bonding between negatively and positively charged ions is known as **ionic bonding**.) The solid aluminium hydroxide then sinks down to the bottom of the lake. Although this makes the water safer for animal and plant life, it can poison the organisms living at the bottom of the lake. The other problem is that after the metals become concentrated in the bottom layers they are able to dissolve back into the water later on when liming stops. Liming is therefore only a temporary method of preventing or reducing acidification.

The most obvious, and best, way of lessening the environmental damage caused by acid rain is to reduce the amount of sulphur and nitrogen gases being released into the atmosphere. Power stations and factories can have a special fitting attached to their chimney stacks to absorb and treat the sulphur and nitrogen waste gases before they are released. However, these fittings add to the cost of construction and can lead to inefficiency, so the cost to industry has to be weighed against the damage to the environment.

Figure 3, *below left: Lime being dropped by helicopter;*

Figure 4, *below right: Dying trees*

Pollution of rivers

What are dead rivers?

Most people have come across small rivers or streams which smelt bad, looked black or murky and showed obvious signs of pollution such as rubbish, or even dead fish, floating on the surface. These rivers or streams can be described as **dead** or **dying**. In the nineteenth century, the River Thames became so polluted that a meeting of Parliament actually had to be stopped because of the terrible smell from the river.

Figure 5, *left: Dead fish in a polluted river*

Box 2   Poor and bad quality rivers

- The 1985 River Quality Survey classified 11 per cent of rivers in England and Wales as of either **poor** or **bad quality**.
- Poor quality rivers are so polluted that they usually don't contain fish.
- Bad quality rivers are very badly polluted and are likely to upset people by their smell and appearance.

Dead rivers are those in which most natural forms of life have disappeared because of pollution. River pollution destroys life in many different ways; one way is by introducing harmful chemicals such as cyanide, ammonia and metals. However the most important factor is the loss of **dissolved oxygen** in the water. Without enough dissolved oxygen, a river cannot sustain the life of the various organisms which are the outward signs of its health.

How do healthy rivers deal with pollution?

A healthy river can absorb and purify a moderate amount of pollution without any harm being done to the aquatic life.

Box 3   Purification

The process of purification involves:

- settling of solid materials and particles to the river bed
- death of harmful bacteria and other organisms due to the inhospitable water conditions
- **biodegradation** of organic waste – the most important effect

Much of the pollution in rivers is caused by the constant discharge of water from industry and sewage treatment works. These water flows contain large amounts of **biodegradable** material. (Biodegradable materials are those that can be broken down to simple chemical compounds by natural organisms.) When a polluted flow enters a healthy river the organisms feed on the pollution and, as they do, their numbers increase. Because there are more organisms requiring oxygen, the amount of dissolved oxygen in the river falls. The polluted flow therefore causes an **oxygen demand** on the river. This oxygen demand (in milligrams of oxygen per litre of water) is called the **biochemical oxygen demand**.

How do rivers die? Rivers die when there is so much pollution that there is very little, or no, dissolved oxygen in the water. So the amount of dissolved oxygen must be kept as high as possible.

Box 4 Dissolved oxygen

The amount of dissolved oxygen in a river is the balance between the loss of oxygen (used up mainly by organisms) and oxygen received naturally from the atmosphere. If the dissolved oxygen in the river falls below its highest natural level (the saturation concentration), it will naturally dissolve more oxygen from the atmosphere in order to reach saturation again.

Cold, fast-flowing mountain streams appear to be wonderfully clear and clean, and this is partly because the speed of the stream causes turbulence which can easily draw oxygen from the atmosphere into the water. The temperature is also very important as water can hold more dissolved oxygen at lower temperatures. At the opposite extreme, in a stagnant pond on a hot summer's day, the water is still and warm, and therefore less able to receive and retain oxygen from the atmosphere.

In most flowing rivers the actual amount of dissolved oxygen is less than its maximum (or saturation) level since there is usually some organic matter being purified, even in rivers which are only slightly polluted. When a large amount of pollution enters a flowing river the level of dissolved oxygen falls very quickly. However, the bigger the difference between the actual amount of oxygen and the saturation value, the faster the oxygen is drawn into the water from the atmosphere. As the river moves on, away from the point of pollution, the dissolved oxygen will therefore fall to a minimum value and then rise again. If the minimum value is not too low then no harm will be done. However, if the pollution is very strong then the minimum value may reach zero, and stay at zero for a considerable length of the river, causing it to die.

POLLUTION

How can river pollution be controlled?

In the past, Acts of Parliament have made it illegal to discharge any polluting flow into a river deliberately without a licence from the local water authority. By these licences the water authority can control the amount of pollution, and where it enters, in order to keep rivers healthy.

Most rivers, where the amount of dissolved oxygen is more than 60 per cent of the saturation amount, can absorb and purify pollution equal to a biochemical oxygen demand of 4 milligrams of oxygen per litre without damage being done to the life of the river. The maximum allowable biochemical oxygen demand of the polluting flow will depend on the dilution when it enters the river. Twenty milligrams per litre is the limit which was used by the water authorities for many years as the necessary standard for treated sewage flows from sewage works. This assumes that each litre of treated sewage is diluted by 8 litres of river water, and if the dilution is less, the water authorities could set a lower limit on the biochemical oxygen demand of the polluting flow.

Pollution of bathing beaches

In the UK we depend very much on the sea for the treatment and final disposal of sewage from our coastal areas. In fact, about 30 per cent of our total sewage flow is discharged directly by pipeline to the sea after only a small amount of treatment.

Figure 6, *below: A coastal sewage pipeline at a bathing beach*

Why are some of our beaches polluted?

Most of the coastal sewage pipelines, or **sea outfalls**, were designed and constructed at a time when we were less worried about environmental pollution than we are today. It was thought that the sea would dilute the sewage flow so quickly, and by so much, that there would be no visible sign of pollution and no possible danger to the public. However, many of the sea outfalls near bathing beaches are so short that the sewage is not properly diluted and taken out to sea. Instead, it is quickly returned to the shore by the tide, or on-shore currents, and this results in pollution of the local beaches.

We can get some idea of the extent of this type of pollution from a 1986 survey which tested bathing waters throughout England, Wales and Northern Ireland. Preliminary results show that the quality of over one-third of our bathing waters failed to meet international standards.

How do we define a polluted beach?

The quality of the sea at bathing beaches is measured mainly in terms of the number and type of **micro-organisms** in the water, and any visible material that has come from the sewage. Micro-organisms are tiny living organisms (such as bacteria, fungi, viruses and algae) which can only be seen under a microscope. The number of these micro-organisms must be kept low in order not to be a health hazard, and there must be no obvious sewage particles or organic matter in the water.

Box 5   International standards for bathing water

Although the international standards include several types of micro-organism, the faecal coliform bacteria is particularly important, and is widely used as an indicator of human pollution. The maximum number of faecal coliform bacteria permitted in bathing water is 2000 per 100 millilitres of water.

The international standard of 2000 faecal coliforms per 100 millilitres is in fact very low compared with the 10 to 100 million per 100 millilitres which is typical when untreated sewage is discharged into the sea. Consequently, if any of the sewage pollution does return to the shore later, the number of micro-organisms must have been

Is it safe to discharge sewage into the sea?

reduced by between 5000 and 50 000 times to conform with the international standard.

The safest way to discharge sewage into the sea is to run a sewage pipeline so far from the shore that none of the sewage flow can return to the shore, and this type of pipeline is known as a **long sea outfall**. Because long sea outfalls are very expensive to construct, most of today's outfalls are only long enough to ensure that the end of the pipe is still submerged at low tide.

How is sewage diluted?

When sewage is released at the end of a submerged pipeline it rises upwards because sewage is lighter than seawater. If the sewage is warmer than the sea, which is usually the case in the winter months, this difference will be greater. As the sewage rises, it mixes with the surrounding seawater and becomes more and more diluted. The final dilution of the sewage when it reaches the surface of the sea depends on the speed of the sewage leaving the pipeline, the diameter of the pipe and the depth of water above the outfall. It is recommended that the sewage should be diluted 100 times before it reaches the surface to prevent a visible sewage layer (or slick) forming.

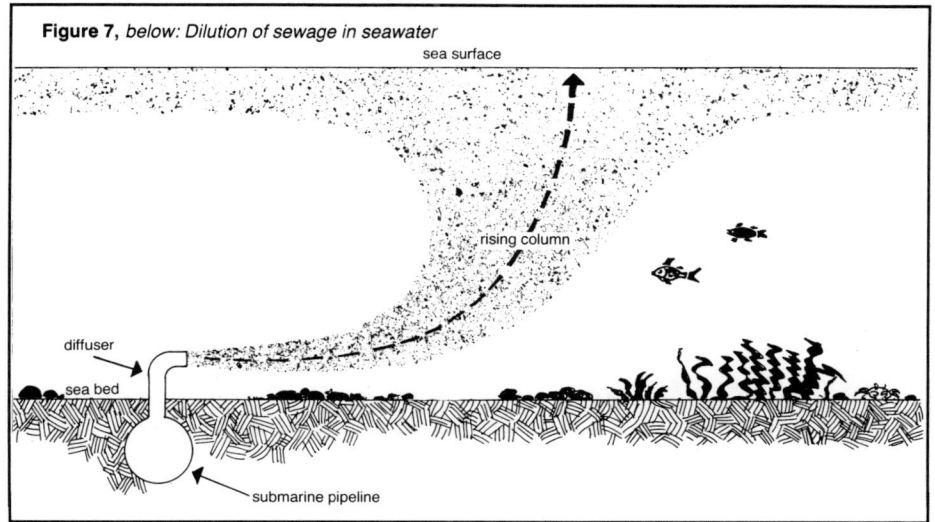

Figure 7, *below: Dilution of sewage in seawater*

How is sewage dispersed? When the diluted sewage reaches the surface, it spreads outwards forming a layer about 1 metre thick, often in the shape of an ellipse. It is then dispersed by the diluted sewage mixing with seawater – and this happens both vertically and horizontally. The amount of vertical mixing is affected by differences in water temperature and density, the speed of sea currents, the roughness of the sea bottom and the effects of wind and waves. The amount of horizontal mixing can also differ according to the effects of tides and other local sea currents.

Box 6 Observing horizontal dispersion

The horizontal dispersion of sewage can be traced by watching the movement of special buoys or floats, which give useful information on the speed and direction of sewage discharges at particular times of the year and under certain wind and tide conditions. Dispersion also reduces the numbers of sewage micro-organisms, and this can be predicted by releasing a tracer chemical (often a fluorescent dye) at the approximate position of the sewage discharge and measuring the amount of the chemical in samples taken from the bathing water.

How long do sewage micro-organisms survive? Apart from the effects of dilution and dispersion, the numbers of sewage micro-organisms are also reduced naturally as they die off. The sea is a hostile environment for them because of its low temperature, salt content and lack of food. However, the main cause of micro-organism death is thought to be exposure to ultraviolet radiation in daylight.

The more the sewage is diluted as it reaches the surface of the sea the less 'cloudy' will be the mixture of sewage and seawater. A clearer mixture will allow more daylight to penetrate the sewage and cause more micro-organism death. This is another reason why the sewage should be more diluted.

Micro-organism death is measured by the time taken for the number of organisms to reduce by 90 per cent, and although it is very approximate, a time of 4 hours is fairly typical. In theory this means that if, after the initial dilution, there are 1 million faecal coliform bacteria entering the sea per 100 millilitres and if we ignore the effects of dispersion, the population will reduce to 100 000 after 4 hours, to 10 000 after 8 hours and 1000 after 12 hours.

How can sewage disposal be made safer and more efficient?

To make sea disposal of sewage safer and more efficient, we need to have a good knowledge of the movement of the sea throughout the year at the point of discharge. This knowledge allows us to decide on the best position for the end of the outfall, how long the pipe should be, and at what times of the day sewage can safely be discharged. Because many outfalls are not discharging sewage safely, water authorities may in the future have to rely on first adding chemical disinfectants to the sewage to destroy the micro-organisms.

Pollution of tap water

How pure is our tap water?

Most people realise that although tap water is usually clean and safe it is not pure water; it almost always contains small and harmless amounts of **organic** and **inorganic substances.** (Organic substances are those that contain the element carbon.)

Of the many inorganic substances in tap water, lead and nitrates are now causing concern because they have been found in some water supplies at levels of concentration above international safety standards. Lead and nitrates occur in different ways so the methods used to control their presence in tap water are also different.

How much lead does our tap water contain?

Lead exists in the environment in the form of chemical compounds (e.g. lead carbonate) or as a positively charged ion. We are constantly exposed to it in food, water and air, and a typical intake is 1.4 milligrams per week, of which around 70 per cent comes from food, 10 per cent from air and 20 per cent from water. Only a small fraction of this amount is absorbed by the human body, so a typical adult will absorb about 0.15 milligrams per week. In order to avoid damage to health, the World Health Organisation has suggested that the total intake of lead from all sources should not be greater than 3 milligrams per week per person.

In general, the concentration of lead in natural

waters (such as rivers) is low, and the usual concentration in British drinking water is approximately 0.02 milligrams per litre. However, in areas of the country where the plumbing system is based on lead pipes the concentration of lead can be much higher, and may occasionally be more than 1 milligram per litre. At this concentration, the intake of lead from water alone would be well above the safe level recommended by the World Health Organisation.

Where does the lead come from?

The major sources of dissolved lead in our drinking water are lead pipes and lead jointing material used in the water distribution network. This process of lead dissolving in water is known as **plumbosolvency**, and the amount of dissolved lead is much greater in acidic or **soft waters**. A soft water is one where the natural amount of calcium and magnesium in the water is relatively low.

What can be done to reduce the lead content?

Obviously, the only way to guarantee low levels of lead in our drinking water is to replace all lead pipes with safer materials, but since there are several million households with lead pipes or fittings this is not practical. An alternative is to reduce the plumbosolvency by making the water slightly alkaline (or **hard**) and water authorities do this by adding lime and soda ash (sodium carbonate) at the water treatment works. These chemicals cause the formation of a solid **precipitate** of lead and calcium carbonate, rather like the 'fur' which builds up in a kettle. This coats the inside surfaces of the pipes and fittings as the water passes through, and stops any further dissolving of lead.

How much nitrate does our tap water contain?

Nitrates are chemical compounds of a metal, nitrogen and oxygen which exist in water as a positively charged metal ion and a negatively charged nitrate ion. Because they are not removed during normal water treatment and supply, concentrations of nitrates in tap water will be similar to those in the original water source. There is some evidence that drinking water with a high concentration of nitrates can cause illness, particularly in babies, so an upper limit of 50 milligrams per litre is recommended.

POLLUTION

Where do the nitrates come from?

There are two main sources of nitrates in river water. Firstly, they are formed from the biodegradation of some organic waste during normal sewage treatment. Treated sewage therefore contains a considerable amount of nitrates when it enters the river. The second source is the drainage of agricultural land. Nitrates are contained in most soil fertilisers because they are a very important nutrient for crops and plants. When it rains, some of the fertiliser on the land is washed into nearby streams and rivers. In some cases, this water – containing nitrates – percolates down to contaminate underground waters which are being used as drinking water supplies.

What can be done to reduce the nitrate content?

The nitrates formed from the biodegradation of organic waste during sewage treatment can be removed at the sewage works by a special biological process. Nitrates from agricultural drainage can only be removed at the water treatment works – before the water goes into public supply – and this involves the same biological process. This process is called **biological denitrification**.

Biological denitrification uses bacteria to break down the nitrates. To do this, the bacteria also require carbon which is present in any organic substance. Sewage already contains enough biodegradable organic matter to support the bacteria. River water, however, must have an extra source of organic carbon, such as methanol, added to it before denitrification can occur. In simple terms, both the methanol and the nitrates are converted by the bacteria to carbon dioxide and nitrogen, which are released into the atmosphere.

Conclusion

In this chapter we have seen how Man and the environment can be affected, directly and indirectly, by persistent water pollution. It is clear that the pollution of freshwater rivers and lakes, and coastal beaches, is the result of the continuous and indiscriminate release of Man's human and industrial waste into the environment. As our knowledge and awareness increase, it is to be hoped that we will all realise the importance of reducing these damaging effects – both in our own interests and in the interests of future generations.

5 LASERS
The light fantastic

Robert Eason

> **This chapter investigates:**
> - what lasers are
> - how lasers work
> - different types of lasers
> - how lasers can be used
> - laser 'weapons'
> - lasers in the future

This chapter is about some remarkable devices which, although they have only been around for less than 30 years, have had an almost revolutionary effect on a wide range of areas including scientific research, medicine, telecommunications and the entertainment industry.

The use of tiny lasers in compact disc players, and laser barcode-readers at supermarket check-outs are examples of fairly sophisticated consumer products which we take largely for granted. However, the hand-held laser gun, that vital accessory for all self-respecting science fiction heroes, is not available so far – yet most children and quite a few adults probably think it is, and are somehow disappointed when scientists fail to produce one. The public reaction to the 'Star Wars' idea of an all-embracing safety net based on lasers in space is similarly unrealistic. Despite overwhelming technical problems, many people believe that such a system is just round the corner.

What are lasers?

Lasers produce beams of light which are rather different to light from other more familiar sources such as the sun, or an electric light bulb. Laser light is a very 'pure' form of light which has several distinguishing characteristics.

Laser light is highly directional

Unlike ordinary light, for example from a torch, which tends to spread out into a broad cone, laser light is produced in a highly directional or concentrated form. The technical term is **collimated** light, and means that a beam of light with almost parallel edges comes out of a laser. This collimated beam spreads out very slowly as the light travels away from the laser, so that even if you shone a laser beam at the moon, for example, it would only have grown to a few square kilometres by the time it reached the lunar surface. If we compare this to another form of less collimated light, such as a very powerful searchlight, we find that the beam spreads out much more rapidly, and fades as it travels.

Laser light is of a single colour

A laser beam is 'pure' because the light is only a single colour. If you took white light from a light bulb, say, and passed it through a glass prism, you would see a whole spectrum of colours (the familiar rainbow effect sometimes seen in corners of mirrors, or cut glass). However if you then looked at the red part of this spectrum on its own (as in *Figure 1*), you would find that it was actually composed of a smaller spectrum of many closely spaced frequencies or wavelengths. This so-called 'red' light is therefore still not very pure.

By contrast, lasers produce **monochromatic** or single-colour light; each type of laser generates one, or several, separate 'pure' colours.

Figure 1, *below: The electromagnetic spectrum*

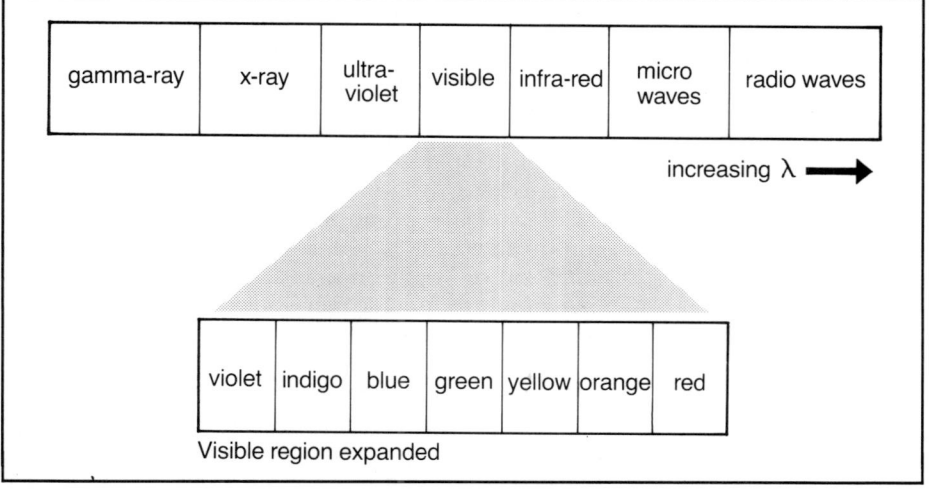

Box 1 Light: Wavelengths and the visible spectrum

Visible light is that part of the **electromagnetic spectrum** that our eyes are sensitive to. There are also light waves outside this rather limited region, and they can fall into other categories such as x-ray, ultraviolet, infra-red or even radio frequencies (*see Figure 1*).

We can work out where light of a particular sort falls in such a spectrum by measuring its **wavelength** (λ), which is the distance between two similar portions of the light wave (such as two successive peaks or troughs). A snapshot of any wave would look like *Figure 2*, and we could easily measure its wavelength in centimetres, for example.

In the case of water waves produced by dropping pebbles in a pond, the wavelength might be a few tens of centimetres. Light waves in the visible region of the spectrum, however, have a wavelength of between 400 nanometres for blue light and roughly 750 nanometres for deep red light.

1 nanometre (nm) = 1 thousand-millionth of a metre or 10^{-9} m

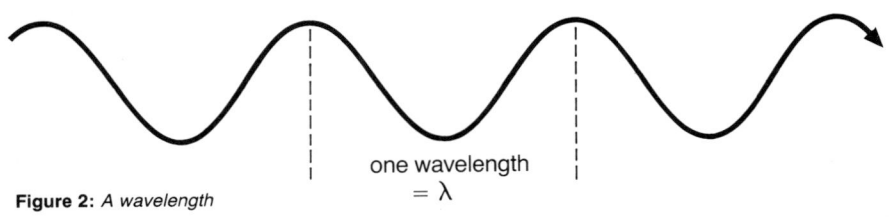

Figure 2: *A wavelength*   one wavelength = λ

Laser light waves are 'in step'

The third important feature of laser light is its **coherence**, which means that the light waves are all 'in step' with one another. It's rather like dropping a pebble into a pond. If you took a handful of pebbles and dropped them one by one in a random way, the ripple patterns produced would all have roughly the same characteristics (same height, same wavelength) but the individual patterns would not be co-ordinated with each other in any way. Each wave pattern would be independent or 'out of step'.

Likewise, if a group of soldiers has no leader or commander calling 'left, right, left, right', they have no means of synchronising their steps. And the marching group looks very odd with each member being essentially independent of his neighbour.

In both these situations there is no co-operative behaviour, and neither the wave patterns produced by the pebbles, nor the soldiers marching, are in step at all. The laser word for being in step is in **phase**. In laser light, part of the beam is in step or in

> **Box 2   Light: Wavelength or frequency**
>
> A different but equally valid way of describing the colour of light is by its **frequency (f)**. If we watched the water ripples passing a certain point in the pond, we might count, for example, five waves in one second. This is written as 5 **cycles** per second or 5 **Hz** (named after a scientist called Hertz). Because light waves have such a small wavelength, and move so fast, their frequency is very high. Blue light, for example, has a frequency of nearly 10^{15} Hz, or one thousand-million million cycles per second.
>
> Frequency (f) multiplied by wavelength (λ) equals the **wave velocity**, or speed (**c**). This statement can also be expressed as
>
> $$c = f \times \lambda.$$
>
> Frequency and wavelength are examples of the sometimes confusing way scientists choose to describe the same thing, using two different types of measurement.

phase with every other part (at least over a certain distance or time). The reason for this will become clear in the section on how lasers work. In a sense, each part of the light beam is pulled into phase by a type of chain reaction. The moment a laser is turned on, the light it produces triggers off more light with exactly the correct phase, which in turn produces more light, and so on. The resulting light waves all have their peaks and troughs at the same place in the light beam, and each part of the beam behaves in a very similar way to every other part.

Light which is locked together in phase is called coherent light, and is produced by lasers. All non-laser sources, such as the torch in *Figure 3*, produce incoherent light. This means that every part of the

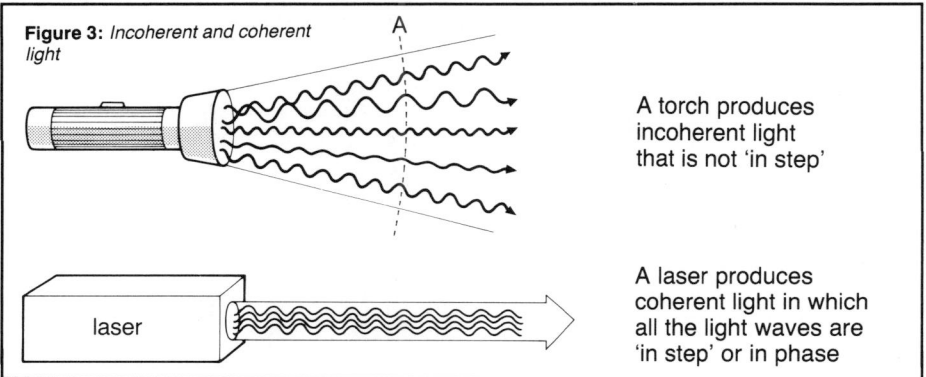

Figure 3: *Incoherent and coherent light*

A torch produces incoherent light that is not 'in step'

A laser produces coherent light in which all the light waves are 'in step' or in phase light beam is different. Although, for example, the two top waves from the torch are in step at position A, they have different wavelengths and may therefore never be in exact step at any other place in the beam.

Laser light is very bright

Because lasers are collimated, monochromatic and coherent, they can produce a beam of light that is very bright to look at. In fact looking directly *at* a laser beam is a very dangerous thing to do. Even a comparatively weak beam of collimated laser light can be powerful enough to hurt your eye seriously, or even cause permanent damage and blindness. The brightness of a beam of light can be defined firstly by its **power** measured in watts (think of a 60-watt light bulb) or secondly by the **intensity** which refers to the amount of power falling on a certain area, for example 1 square centimetre.

Some lasers have enormous power. For example, lasers used to cut materials like sheet steel can have a power of 10 000 watts, which can be focused down to very small sizes, to produce intensities of perhaps 1 million watts (or 1 megawatt) per square centimetre. These lasers are used for industrial cutting and welding because even steel is rapidly melted at this intensity.

How do lasers work?

What's inside a laser?

The word laser is actually short for **L**ight **A**mplification by **S**timulated **E**mission of **R**adiation. We already know what light is, but what is meant by **amplification** and **stimulated radiation**?

Figure 4 should help to explain these terms because it shows what most lasers look like in principle. Although the picture illustrates a ruby laser (which uses a rather expensive and precisely made crystal of synthetic ruby) the actual laser material doesn't matter at this point. The ruby crystal is a solid, but the laser would work equally well if a liquid, gas or vapour were used as the active laser material instead. This ruby laser was actually the first one to be invented – by an American scientist called Theodore Maiman in 1960 – and ruby lasers are still in use today. Since Maiman's remarkable

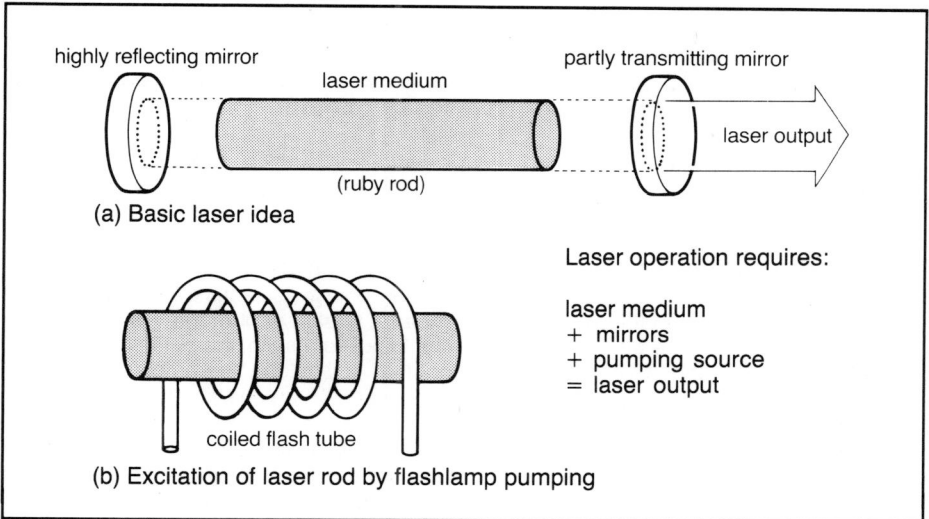

Figure 4, *above: A ruby laser*

discovery, the past 28 years have seen an explosion of interest in lasers and laser-related research. Today lasers can provide a very wide range of light frequencies, spanning infra-red to visible, and even the x-ray region of the spectrum (*see Figure 1*).

If you looked inside the ruby rod in *Figure 4*, using a sufficiently powerful microscope, you would see that it was made up of atoms, just like all substances. In the case of ruby, only one sort of atom – a chromium atom – is of interest to us, and chromium makes up roughly 0.1 per cent of the ruby rod. These chromium atoms are responsible for the lasing action (i.e. producing the laser output) but first they have to be excited from their normal level of inactivity into an active state. This is done by a method called **pumping** in which energy is put into the lasing atoms by an incoherent light source, such as a flashlamp, so that they can re-generate this energy in the form of a coherent, monochromatic, directional laser beam.

Pumping is not as strange an idea as you might think. **Fluorescent** strip lights, for example, contain gases that are excited by collisions with electrons. The excited gas atoms are energy-rich, and after a short space of time, perhaps a millionth of a second or so after being excited, they give up this newly acquired energy and emit a small particle of light called a **photon**.

Box 3 Light: waves or photons?

Why do scientists often use the terms 'light wave' and 'photon' almost interchangeably? At first glance this seems rather curious for people who are generally so organised and precise.

A beam of light is made up of a stream of photons which collectively have wave-like properties, but when studied individually might seem to behave more like particles. So when dealing with light and light energy, scientists tend to talk about light waves even though they really mean a stream of photons. It's rather like drawing a line on a piece of paper. It may look like a solid line but in fact it's made up of thousands of dots very close together. If this idea seems confusing, don't worry. It worries most scientists at some level – 'Everyone *knows* what light is, but few can *say* what it is.'

The area of physics which discusses this rather perplexing problem is quantum mechanics. The concepts of particle and wave were combined by people such as Einstein, Heisenberg, Dirac and others into an elegant new theory which described the way microscopic particles such as electrons and photons behave. Quantum mechanics has been very successful in predicting new effects and discoveries, and has revolutionised 20th-century thinking about light, matter and particles.

Laser amplification

In a strip light all the emitting atoms are not behaving in a co-operative way – each atom is working by itself. Something else is required to make it a laser, and the missing link is amplification.

Let's look at *Figure 5*. Stage 1 shows the ruby rod in its normal unexcited state. After the flashlamp has been fired, and the atoms inside the rod have been excited (stage 2), fluorescence starts to occur, and atoms emit their photons in different directions (stage 3).

However the photons that happen to travel along the rod and are reflected by one of the mirrors can trigger off other excited atoms to emit their photons. The mirrors therefore provide **feedback**, which is vital for a laser. The process of coaxing an atom to emit its photon before it does so of its own accord is called stimulating it (stage 4). The stimulating photon produces a second stimulated photon. The two photons are identical in their direction, their phase and their colour.

If the conditions are right (stage 5), this stimulating process continues, using the mirrors constantly to reflect light along the rod, until stage 6, in which the laser is fully on, and an avalanche of photons is produced, all in a coherent state.

Figure 5, *below: Laser action*

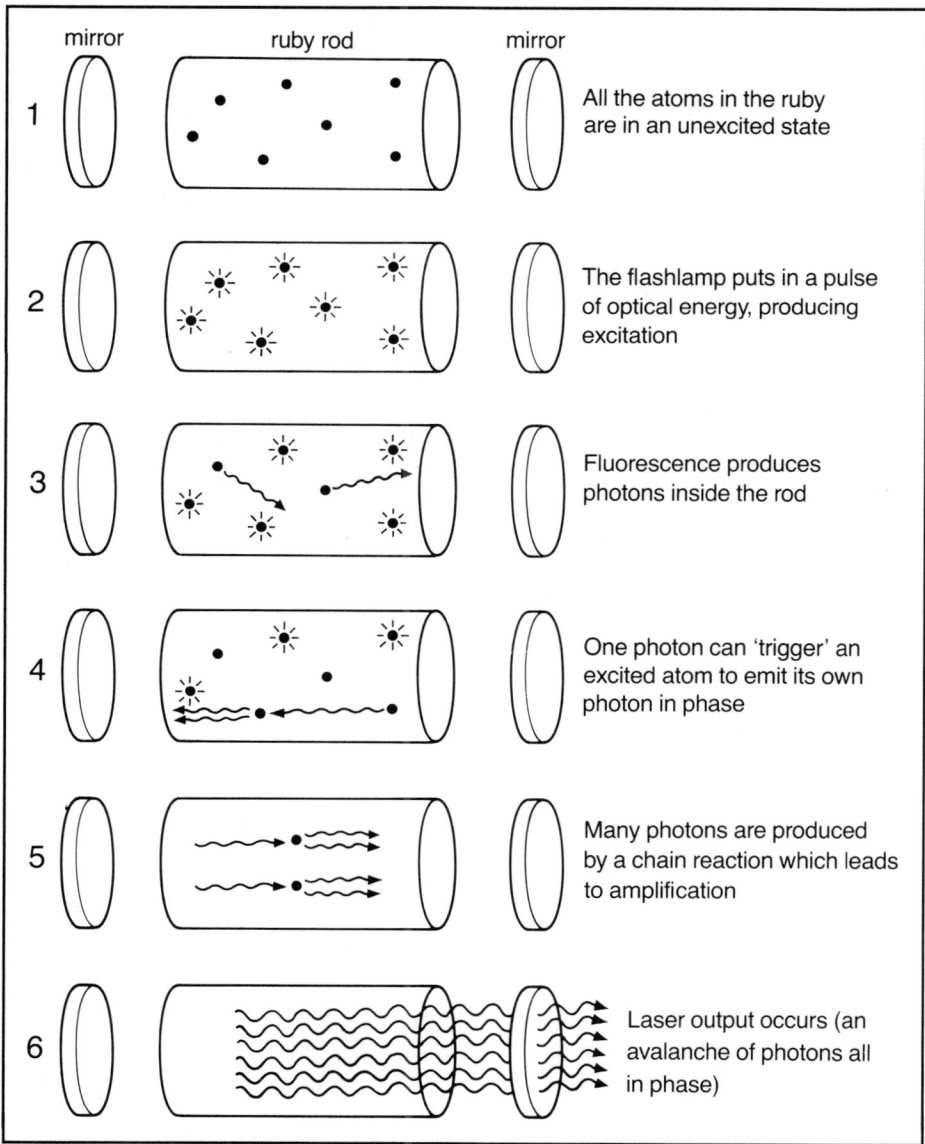

It is clear therefore that lasers require mirrors for amplification. However, if the photons are constantly reflected back by the mirrors you may wonder how the laser light ever gets out of the cavity. The answer is that the front mirror is made 'leaky' or only partially reflective, so that some of the laser light can escape and produce a useable laser output. In practical lasers the reflectivity of the back mirror is usually close to 100 per cent, while the front mirror might have a reflectivity in the range of 50–95 per cent, for example.

Types of lasers in common use

There are several ways of classifying the many different types of laser. Perhaps the best way is to take several examples of lasers used in a range of applications, and to describe the way in which their characteristics are matched to the job requirements.

Industrial lasers

The industrial uses of lasers mainly involve cutting, welding and drilling of materials which can range from sheet steel to plastics, as well as the harder (and more precious) materials such as diamond. Industrial lasers are often seen on the production line, where they can operate 24 hours a day in a computer-controlled environment, performing extremely precise and intricate machining operations. Unlike conventional techniques, lasers never require a halt in production to replace a worn drill bit or renew a welding rod.

For welding, a continuous laser beam is often used, at power levels greater than 1 kW (1000 watts). The type most commonly used is a carbon dioxide (CO_2) laser in which the active laser material is simply CO_2 gas contained in a suitable laser tube. Laser welding is relatively straightforward – the laser beam is simply moved across the join between the two pieces. The laser only heats and melts the metal next to the join, so the weld is very precise and well controlled. Sheet steel up to 2.5 cm thick can be welded at a rate of 2 metres per second, using large CO_2 lasers.

Drilling by laser is similarly straightforward. A suitable laser is directed at the point where the hole is required, and the material is then removed by vaporisation. For drilling, pulsed lasers are often used because they emit short bursts of laser energy, so the material is vaporised a little at a time. Laser-drilled holes can be very precise, and also much smaller than those produced by conventional mechanical drills.

Medical use of lasers   The cutting and drilling capabilities of lasers need not only be applied to the inanimate world of steel, plastic or glass. Lasers are also used in hospitals on most parts of the human body, particularly the skin, brain and internal organs such as the stomach. They provide an alternative tool for the surgeon who may reject the conventional scalpel when faced with a delicate operation on, for example, the eye, or an operation in an inaccessible place such as the stomach or cervix.

Medical lasers have many advantages. Firstly, the laser beam is pure energy, and achieves the cutting or heating without any physical contact. This means that the chances of infection (as from an infected scalpel) are nil, and the heat from the laser sterilises and seals blood vessels as it goes.

Another advantage of lasers relies on the use of **optical fibres** – very thin strands of glass, usually – which can transmit laser light to inaccessible places within the body. Instead of opening up the stomach or throat, for example, a bundle of optical fibres can easily be inserted into the correct area (by, for example, using a syringe) and laser light directed at the affected part – perhaps an ulcer or a tumour.

A third benefit is that the colour of the laser can be finely adjusted so that only the affected part absorbs the laser light. A dye laser (which uses a chemical dye in water or alcohol as the laser medium) can be made to operate at any wavelength in the visible spectrum, so that a red birth mark, or deep red 'port wine stain' can be selectively burned away, leaving healthy 'pink' skin nearby unharmed.

Laser eye surgery has been particularly successful. An operation known as 'retinal welding' is routinely performed for patients who suffer loss of clarity in their vision caused by detachment of their **retinas**. By shining laser light at the patient's eye, the lens in the eye focuses the laser beam on to the retina, and tacks the retina back in place with a spot-weld.

Holography A **hologram** is a kind of photograph which is produced when laser light reflected from an object is made to interfere with a secondary or reference beam of light derived from the same laser. The special pattern produced on the film or plate records not only the intensity of the reflected light, as in an ordinary photograph, but also its phase.

Interference of light beams can be likened to the interference of waves in water. If the circular ripple patterns, produced when two pebbles are dropped in a pond, overlap, then points can be seen where two peaks coincide, and the resulting pattern appears twice as big. These are known as regions of constructive interference. Similar points of destructive interference can be seen where a peak from one pattern overlaps a trough from the other, and the resultant is zero. These interference or ripple patterns produced by mixing the light from the object and the light from the reference beam are recorded on the hologram, and give us a unique record of what the light from the object 'looked like'. This first step – the formation and chemical developing of the holographic pattern – is called the recording process (*see Figure 6a*).

The second step – the **reconstruction** (*see Figure 6b*) – is the part that gives holography its universal appeal. When the holographic interference pattern is re-illuminated with laser light of the same wavelength and from the same position, an image is formed in the same place as the object, with all the three-dimensional clarity of the original. The image seems to hang in space, and appears so lifelike, that the temptation to 'touch' it is overwhelming.

The holographic reconstruction, observed by looking through the hologram as if it were a window, has the properties of **parallax**, which means that you can examine the top, bottom, and sides of the image by looking at it from different angles, just like the original. A variation on the reconstruction process in which the hologram is illuminated from the opposite side, produces real images which appear to loom out of the plate and hang in space before your very eyes.

Figure 6, *below: Holography*

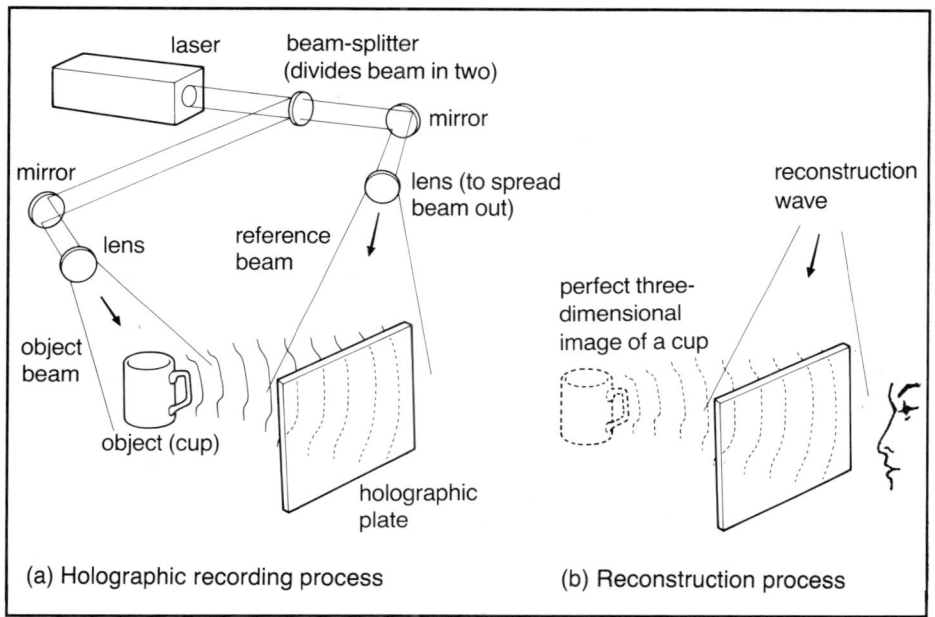

(a) Holographic recording process

(b) Reconstruction process

The incredible realism of holographic images really has to be seen, rather than described second-hand. Exciting variations on the holographic principle now include reflection holograms which need a laser to make them but can be reconstructed with ordinary white light. Other kinds of hologram use media other than holographic film, such as special crystals, which produce **dynamic** (or moving) holography. If you need evidence of the widespread use of holography today, take a look at the holographic logo on the bottom right-hand corner of a banker's card or credit card.

Lasers for entertainment

Lasers have also found extensive application in the entertainment world. There are light shows where laser beams are rapidly scanned across large screens to create dramatic and complex multicoloured pictures, which can be made to move or pulsate in time with the music. Sky-writing, where laser light is shone at clouds and the beam moved to write messages, logos or adverts, is used for advertising.

Perhaps the most common use of lasers for entertainment, however, is for compact disc (or CD) players, which many people now own. Light from incredibly small **semiconductor lasers** (much smaller than a pinhead) is very precisely directed and focused on to the surface of a disc which has been encoded with music as a series of dots or 'pits'. The method of recording on to these discs is very different to the more familiar technique used for conventional records.

Compact discs use **digital** recording, rather than the **analogue** recording used on an LP record. The music is recorded in the studio in a code which only has on and off signals, rather than a continuous range of sounds and notes. The on/off code is then printed on to the CD surface as a series of indentations or pits.

This idea of using on/off signals rather than continuous sounds is not particularly strange. Morse code, for instance, uses a similar technique, and all computers work on a system of on/off signals called **binary** operation.

The CD has a reflective metallic surface covered by a protective coating of clear plastic. The player spins the disc, and the laser scans tracks recorded across the disc surface. If the light hits an area of the disc with no pits then it is reflected back, and the electronics register an 'on' signal. A pit, however, will not reflect the light, so an 'off' is produced. The CD player then decodes this rapidly varying on/off signal back into the original music.

The great advantage of CD recording is that the problem of noise (hiss and crackle) is completely absent. A CD will play just as well if it has been scratched, and is virtually indestructible, as no physical contact is involved.

Laser weapons: fact or fiction?

The Strategic Defence Initiative (SDI) or 'Star Wars' programme has been in the news a great deal. Since November 1983 when President Reagan made his famous speech about a system that 'could intercept and destroy strategic ballistics missiles before they reach our soil', the world's attention has been focused on the advantages or otherwise of such a scheme. Many people thought an invincible 'space shield' that protected any country from hostile nuclear attack must be a goal worth pursuing.

In fact, long before the famous speech, laser 'weapons' already existed, although in a much more mundane form than that proposed by Mr Reagan. Lasers, with their highly directional properties, have been mounted in place of rifle sights, to show a soldier exactly where his bullet, missile or rocket will land. They have also been used to guide missiles. The missile searches, and then homes in on, scattered laser light which has been directed at a military target by a hidden laser operator. Laser range-finders help soldiers to calculate the exact distance to a target, and hence improve their aiming accuracy.

However these 'weapons' are a far cry from the 'Star Wars' scenario some people believe will benefit us, so we need to look at what is involved in more detail, and try to separate fact from fiction. A report commissioned by the American Physical Society on the feasibility of the SDI program, has recently been published.

Its main conclusion is that the popular view of space-based laser battlestations is distinctly premature. At least ten more years of intensive research are required before we have the technical knowledge to decide whether such weapons can be built.

The report says that four types of laser are currently being considered for SDI. One candidate in particular, a laser which uses reacting chemicals as the laser medium, and which some consider the most promising one, would need to be at least a hundred times better in overall performance. This degree of improvement would require such a huge investment of research and development that it's hard to imagine the technology being available before the next century.

There are other problems however, apart from those of development. The light produced is absorbed by the atmosphere, so it would have to be a space-based rather than a ground-based device. Anyone familiar with the difficulties of operating enormous lasers (some lasers today occupy whole buildings) would doubt the feasibility of this. Other chemical lasers which use different reacting gases might be suitable, but would have to be a hundred thousand times better to stand a chance of success!

The same, or similar, stories apply to the other three laser candidates, which would have to be between a thousand and ten thousand times better than they are now.

The question of reliability, raised in this report, is also an important one. Thousands of these laser battlestations are required to track and destroy the half a million or more 'threat objects' (missiles or decoys) that would be launched in an all-out attack. The largest working laser today, the NOVA laser facility in California, which uses well-established laser technology, requires daily attention to check alignment and cleanliness, as well as regular monthly shutdown periods for maintenance, just to keep it running. This fact contrasts somewhat with the unrealistic SDI picture of unmanned lasers in space, ready to operate at peak efficiency at any time.

However these reasoned arguments do not even begin to answer the associated and equally vital questions of how you track half a million objects simultaneously (so far lasers have only been able to track one object), how you maintain the lasers, how you protect them against hostile attack, how you control and direct them, and so on. There is an informed body of opinion that seriously doubts the credibility of the SDI laser weapons programme, and also believes that the only laser weapons the world will ever know already exist in the form of the range-finders and laser guidance systems described earlier.

Lasers in the future

We have seen how lasers work, what their characteristics are, and described some of their many and varied uses. But one question still remains: what uses will lasers be put to in the year 2001 and beyond?

In the medical field, lasers are really only in their infancy at present. Doctors and surgeons use lasers like rather unsophisticated heaters and saws. The sort of operation that may well become as routine as an eye test today might be, for instance, laser trimming of the lens to correct defective vision. Using optical fibres, a technique called laser angioplasty is currently being tried. In this new treatment, laser light is used to remove the fatty deposits inside clogged arteries. Such medical techniques will undoubtedly be pursued in the near future, so that we might one day be able to forget about healthy eating habits, and simply go in for an arterial clear-out every so often!

In a rather different area, the field of computing is also set to witness a revolution with the introduction of optical computing techniques which are now being researched in, as yet, a fairly modest way. People are talking about computers working with laser light rather than electric currents, and the first all-optical circuits are currently being demonstrated. These all-optical circuits are of great interest to computer scientists and may well prove to be the best method of making computers work in a rather more similar way to the human brain, as we shall see in Chapter 10. The potential for making optical computers which can operate a thousand or even a million times faster than at present seems possible, but only time will tell whether they can realistically be made.

In all aspects of future laser use and development, however, one thing is certain: the laser will never cease to find itself new problems to solve. Lasers will get more powerful, smaller, more efficient and more widespread. Lasers really are a taste of the light fantastic.

6 PLASTIC
Our flexible friend

Ray Smith

> **This chapter investigates:**
> - how a molecule is formed
> - how monomers are combined to form polymers
> - the structure of polymer molecules
> - some everyday plastics
> - manufacturing with plastics
> - composites
> - disposal of plastics
> - plastics in the future

Over the last 50 years plastics have had an enormous impact on our everyday life. These days it's impossible to imagine life without polythene bags, plastic washing-up bowls and non-stick saucepans: 20 years ago it was often insulting to describe something as being made of plastic because it was thought to be a cheap, but less than satisfactory, substitute for natural materials such as wood or metal. Now we often find plastic objects more suited to their purpose than their natural counterparts. This change is largely due to scientists who have increased their knowledge and understanding of these new materials and how they can best be used.

Originally the word plastic was used to describe certain materials – a metal rod is plastic because when you bend the rod it keeps its new shape and does not break. But these days when we talk about a plastic we mean a type of man-made **polymer** or, more often, a polymer to which other substances have been added to give it particular properties.

But what are polymers? The word polymer is derived from the Greek words meaning 'many parts', which gives a clue to its main feature: a typical polymer is a very large molecule formed by joining together lots of smaller **monomer** molecules. And polymers are nothing new. Natural polymers such as leather, wool and cotton have been used for centuries, and it is fair to say that the present-day polymer chemist has a long way to go before he can produce polymers with structures as complex as these natural ones. In a natural polymer many different monomers are combined, whereas a man-made polymer will generally contain only one or two different monomers.

In order to understand polymers we first have to find out about molecules and monomers.

How is a molecule formed?

As we saw in Chapter 1, an atom is made up of protons, neutrons and electrons, and there are 92 different naturally occurring **elements**, such as carbon, hydrogen, oxygen, chlorine, calcium, sulphur, iron and gold. Each element has a different number of protons in its nucleus and together they are the basic building blocks of everything around us.

These atoms can combine together in various ways to form **molecules**, with the possible combinations depending on how the electrons in the atoms are arranged. Electrons can be imagined as orbiting around the nucleus in certain fixed ways, contained in **shells**. An atom of a particular element reacts with other atoms, depending on the arrangement of electrons in their shells. **Chemical reactions** take place in order to fill gaps in the arrangement of electrons, starting from the innermost shell. The structure is rather like an onion: the central core of the onion is in the same position as the nucleus of an atom, and then the layers around the core are like the shells in the atom – each layer has to be there before another one can be added on top.

The first, or inner, shell of an atom is full if it contains two electrons, but *Figure 1* shows a hydrogen atom which has only one electron. The hydrogen will therefore take part in reactions in order to gain another electron to fill that inner shell. Helium already has two electrons, so the inner shell is complete. This stable arrangement of electrons makes it **inert**, which means that helium is unlikely to take part in chemical reactions. The effect of this extra electron in helium can be clearly seen in airship

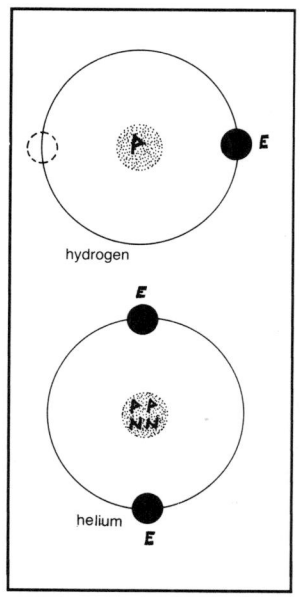

Figure 1, *below: Hydrogen atom and helium atom*

technology where the ship can be filled with either gas. While present-day airships are filled with helium, early airships were filled with hydrogen gas, which is cheaper and lighter. However, because hydrogen has an electron vacancy to fill, it will take part in a reaction.

The Hindenburg disaster was caused by a chemical reaction. It happened because heat provided the conditions for the hydrogen to react with oxygen in the surrounding air to form water vapour. During this reaction more heat was produced (atoms generally give off some energy when they join together to form molecules) and this heat, in turn, helped the reaction to continue. Flames are often the main thing we can see when a reaction is taking place. Each of the water molecules which result from the reaction between hydrogen and oxygen contains two atoms of hydrogen and one of oxygen which have joined together in order to fill their electron vacancies. So why does this happen?

Figure 2 shows an oxygen atom which has a total of eight electrons. The inner shell holds two electrons and is full, so the other six electrons are in the second shell. However the second shell needs eight electrons for it to be complete, so there are two vacancies. This means that oxygen will take part in reactions to fill those vacancies.

Both hydrogen and oxygen therefore need to fill electron vacancies. So when hydrogen burns in oxygen the atoms fill their vacancies by sharing electrons, and they do this by forming **covalent bonds**. Two hydrogen atoms are then firmly fixed to an oxygen atom and a water molecule is formed.

Figure 2, *above: Oxygen atom*

Box 1   Covalent bonding

In the water molecule, each of the two hydrogen atoms shares its electron with the oxygen atom. And, in return, the oxygen allows each of the hydrogen atoms to share one of its electrons. Each of these shared pairs of electrons are, in a sense, orbiting around the nuclei of both atoms. This is shown in *Figure 3*.

PLASTIC

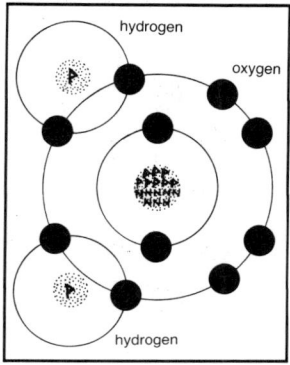

Figure 3, *above: Water molecule*

Molecules are not just mixtures of atoms. The mixture you get from mixing hydrogen and oxygen is still two gases and will show all of the properties of the individual gases. However when they take part in a reaction and form a new compound, water, the properties are quite different. For example, water is not a gas at room temperature; and whereas oxygen tends to help things burn, water is often used to extinguish fires. In the same way that an atom of hydrogen is the smallest possible particle of hydrogen so a molecule of water is the smallest particle of water.

The covalent bonding in the water molecule is only one of the ways in which atoms can be held together to form molecules; the other important way is ionic bonding which was mentioned in Chapter 4. But it is covalent bonding which holds polymer molecules together.

Box 2   Chemical equations

Chemists use chemical equations (or formulae) to show what happens in a reaction, and all the elements have their own 'shorthand' or symbol. Hydrogen is H, oxygen is O and carbon is C. The reaction between hydrogen and oxygen to form water is written as:

$$2H_2 + O_2 \rightarrow 2H_2O$$

This shows that two hydrogen molecules combine with one oxygen molecule to form two water molecules.

How are monomers combined together to form polymers?

To take one example, polythene is a polymer, made from the monomer ethylene, or ethene as it is sometimes known. Usually it is produced from crude oil. *Figure 4* shows that the ethylene monomer molecule (CH_2CH_2) has two atoms of carbon and four of hydrogen, and these are held together with covalent bonds. Carbon has six electrons but requires a full inner shell of two electrons and a second shell of eight, so it has four electron vacancies.

In the ethylene molecule the carbon fills these

Figure 4, *right: Ethene molecule*

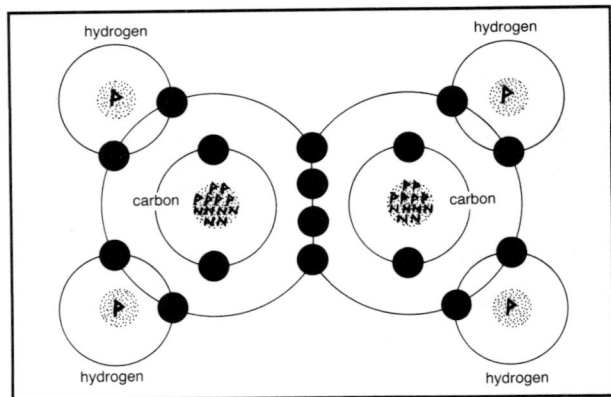

vacancies in two ways: firstly each atom of carbon shares an electron with each of two hydrogen atoms, thus filling two of the vacancies. The carbon atoms then fill their two remaining spaces by sharing two of their electrons with each other, and forming a **double bond**. Each carbon atom therefore shares one of its electrons with each of two hydrogen atoms, and shares two of its electrons with the other carbon atom.

Box 3 Carbon

Carbon atoms have a special property – they are able to bond to other carbon atoms in an infinite number of configurations, and this means that there are an enormous number of carbon compounds. The branch of chemistry which deals with these carbon compounds is known as organic chemistry, and as polymers are compounds of carbon, polymer chemistry is a branch of organic chemistry.

The polythene molecule (also known as polyethylene or polyethene) is formed by breaking the double bond in the ethylene molecule and joining the monomers to each other in a chain, as in *Figure 5*. To get some idea of the length of this polymer: if you think of each monomer as the size of a pearl (1 centimetre in diameter), the chain would be around 200 metres long. The process by which the ethylene molecules are combined to form polythene is known as **polymerisation**.

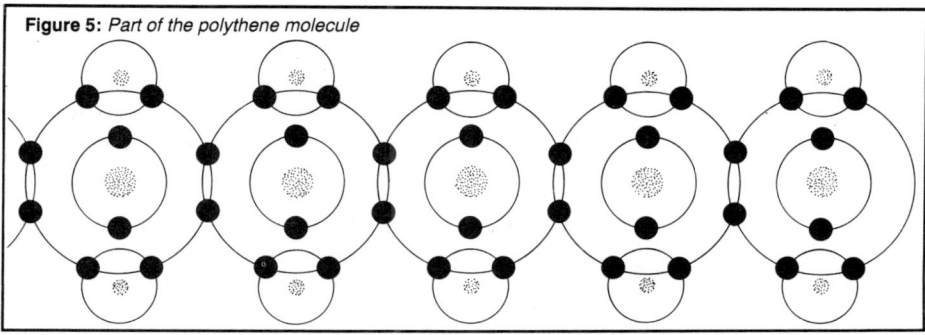

Figure 5: *Part of the polythene molecule*

Making a polymer

Most common polymers are formed from monomers which have been broken down in one of three ways:

- by breaking open double bonds (as in ethylene)
- by breaking a bond in a **carbon ring** (where the carbon atoms are joined together in a circle
- by using monomer molecules which have different reactive groupings of atoms within them.

To polymerise ethylene we need to add energy to break the double bond between the carbon atoms, and this can be supplied in a number of ways:

- ionising radiation (gamma rays)
- ultraviolet light
- heat
- a chemical reaction with a molecule which has an unstable bond and can therefore be readily broken down.

Box 4 The polymerisation equation

The chemical reaction for the formation of polythene is written as:

$$n\,(CH_2 = CH_2) \rightarrow -(CH_2-CH_2)_n-$$

Here the double bond between the carbon atoms in the ethylene molecule on the left-hand side is shown as two long lines. For the polythene molecule, on the right, we show the repeating part of the chain with spare bonds at each end. These bonds are connected to the next part of the chain which is identical; n can be any number and represents the number of monomers making up the polymer.

The structure of polymer molecules

So far we have only discussed polymers as long chain molecules but monomers can in fact be joined together in different ways. Three common structures are illustrated in *Figure 6*. The linear structure can be compared to a pearl necklace, the branched structure to a tree, and the cross-linked arrangement is rather like a net curtain. By varying the polymerisation process, it is often possible to get different structures for the same polymers. In this case two substances, although chemically identical, can have very different properties. This **structure-property relationship** is most important.

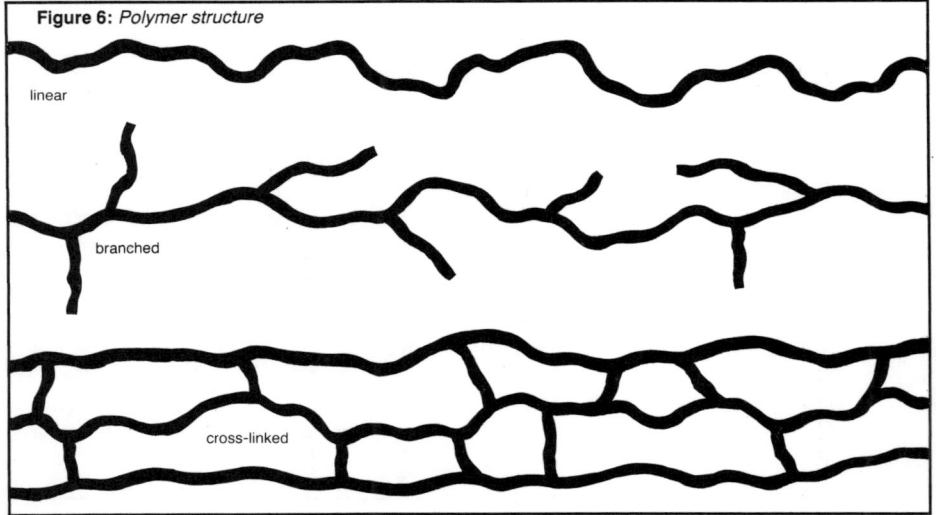

Figure 6: *Polymer structure*

linear

branched

cross-linked

Box 5   Additives

The basic polymers are not always best for the job so, to make a usable plastic, other substances are generally added. Some of the more common additives are:
- anti-oxidants to prevent chemical reaction with oxygen from the air
- stabilisers to limit the way that the polymer breaks up due to heat or ultraviolet light (this is often needed because of the forming process, or for outdoor uses)
- plasticisers to make the polymer more pliable or soft
- pigments to add colour (polymers are usually white or transparent in their natural state)
- fillers to give different mechanical properties (perhaps strength), or to add bulk cheaply (plastics may contain as little as 20 per cent polymer).

Washing-up liquid bottles are a good example of the importance of structure. Until 1966 these bottles were made from low-density polythene whose molecules largely have a branch structure. This means that the molecules are not very close together. Nowadays most of the bottles are made of high-density polythene which has more linear molecules. The new, denser, bottles are more rigid, less squeezy if you like, but as the material is stronger they can be made thinner and so are cheaper.

Some everyday plastics

The plastics that we see in everyday life are really quite few in number, and the main ones are listed in *Box 6*. The two most common ones are polythene and PVC. The reason why PVC is so common is that it is relatively cheap. This shows that a particular plastic is often chosen simply because it is the cheapest rather than the one with the required properties. Manufacturers will often use additives to change the properties of a cheaper plastic instead of employing a more expensive one which may be more suitable.

Although PVC is relatively cheap it is not very

Box 6　Common polymers

Polymer	Some uses
nylon	clothing, ropes, bearings and brushes
polyethyleneterephthalate (PET)	fizzy drink bottles, oven trays and clothing (brand name: Terylene)
polypropylene	bottle caps, washing-up bowls, buckets and kitchen utensils
polystyrene	food containers (some margarine tubs and yoghurt containers); in expanded foam form for ceiling tiles, meat trays and plant pots
polytetrafluoroethylene (PTFE)	non-stick surfaces (brand name: Teflon)
polythene	widely used for packaging (carrier bags, detergent bottles, clingfilm)
polyvinylchloride (PVC)	water and gas pipes, window frames, food containers, clingfilm and squash bottles

flexible, so plasticisers are often added to improve the flexibility, or bendiness. Food containers can be made from PVC with plasticisers and, when used as intended, they are perfectly safe. Problems arise, however, when the container is reused. PVC margarine tubs, for example, should not be put into a freezer because plasticisers may leach out of the tub into the food. This leaching of plasticisers is also the reason why doubts have been raised about using PVC clingfilm for cooking in microwave ovens (incidentally some clingfilm is now made of polythene which does not suffer from this problem, so it's worth checking on the box if you intend to use it in a microwave cooker).

PVC is also often used to make bottles for fruit squashes and mineral water. These days it is often 'orientated PVC', which means that it has been moulded in such a way that the polymer chains all lie in the same direction. This results in a stronger and clearer bottle. PVC cannot be used for fizzy drinks, however, as the fizz (carbon dioxide gas) would simply leak out through the bottle. For this type of drink a different plastic, PET, is used.

PET is also used to make 'dual-ovenable' trays. These trays can withstand temperatures up to 240 degrees Celsius so they can be placed in both conventional and microwave ovens, and this is possible because the molecules are arranged in a crystalline, or highly regular, way.

It's easy to forget that synthetic fibres in our clothing are really plastics and that such fibres now account for almost half the clothing sold in this country. The main fibre used is polyester, which is very closely related to PET.

In PTFE the chains are very smooth so they will not easily attach to anything else. This makes it an ideal material to use for non-stick coatings on saucepans. There is, however, a problem: not only will the PTFE not stick to food, it won't stick to saucepans either! You may remember that early non-stick coatings peeled off very easily; now a mechanical solution to the problem has been found. First, the inside surface of the pan is roughened and coated with a **resin**, which has some particles of PTFE trapped within it. A coating of PTFE is then applied and the whole pan is heated to about 500 degrees Celsius. At this temperature it chemically bonds with the PTFE particles in the resin and the whole coating is secured on the pan.

Manufacturing with plastics

There are two types of polymers and the ways in which they can be used to manufacture objects depends on their type. So far, this chapter has been concerned with **thermoplastics** – plastics which can be formed by heat – which are the most widely used. The other type of plastics are called **thermosetting plastics**.

In a thermosetting plastic, several substances are reacted together, and the resulting mixture is put into a mould. The polymerisation takes place in the mould, chains form and a rigid solid object is produced. Thermosets tend to be more brittle than thermoplastics – probably the best-known one is bakelite. This was the first true synthetic polymer

Figure 7, right: Extruding a multi-layer plastics film

which came into general use in the mid 1920s and is still used today for items such as electrical fittings. It is probably best known as the material used to make telephones for many years.

Thermoplastics become soft at high temperatures – in a non-crystalline plastic, generally between 100 and 200 degrees Celsius – this means that they are easy to mould. A common method is **injection moulding** where molten plastic is forced into a mould at high pressure. Detergent bottles, for instance, are made in this way. Gramophone records are made using another traditional technique – **compression moulding** – where plastic is pressed between two halves of a mould. Other common methods are **extrusion** (which produces pipes, sheets, film or tubes by forcing molten plastic out through a nozzle), **blow moulding** (where molten plastic is blown into a mould) and **vacuum forming** (the opposite of blow moulding where a vacuum is produced in the mould to draw the plastic into the required shape).

Recently, as polymer chemists have learnt more about the structure of polymers, new techniques have been developed. PET fizzy drink bottles, for example, are **stretch blown**. Here the material is stretched in one direction and blown in the other, to give greater strength by arranging the polymer chains in two different distinct directions.

The structure of the plastic can often be changed by the methods of processing. For instance, PET food trays are made from an extruded sheet, and it is during this extrusion process that the material becomes crystalline. The sheet is then vacuum formed to make the tray.

Composites

We mentioned earlier that fillers are often added to a polymer to change its mechanical properties, perhaps to give extra strength or resilience. Many of these filled polymers are known as **composite materials**.

Glass fibre hulls of boats are an example of the use of a composite, in which tiny fibres of glass have been added to a thermosetting plastic to give extra strength. The problem with this sort of material is that the fibres have to be aligned very carefully, otherwise the material will be stronger in some directions than in others. One consequence of this is that such composites are difficult to put into mass production.

Another type of composite material, generally using carbon fibre reinforcement rather than glass, is now widely used in military aircraft because it is much lighter than metal and can be made more rigid. Unfortunately it lacks the ability to absorb energy, and if something hits a carbon fibre aircraft wing it either bounces off or shatters the whole wing. There are fears that the Super Harrier aircraft (which often flies low) could be destroyed by firing a high-velocity bullet at the wing from the ground.

Glass fibre and carbon fibre composites are also frequently used in racing cars and a lot of research is under way into how they could be used in ordinary cars, but so far the problems of impact damage have not been solved. A car with a composite body and, say, an aluminium chassis and engine could be up to 40 per cent lighter than existing equivalent cars and would therefore require much less fuel. Of the mass manufacturers, Citroen have been using composite panels in their cars since 1955. Initially the panels were made from a glass fibre reinforced **epoxy** but more recently they have been using a glass reinforced polyester resin to make the bonnet and tailgate of their BX range of family cars. Using this thermoplastic composite means that they can use injection moulding for mass production.

To be successful however, a family car with a totally plastic body will require panels which are more rigid. The chassis will also need to be strengthened so that the car can withstand the same sort of impact as a car with an energy-absorbing metal body.

Disposal of plastics

Having discussed how you make and use plastics, we also need to know how they can be destroyed. One of the main criticisms of plastics is that they are indestructible. In general this is not true – it is additives which make them so. In PVC, for example, the bonding between carbon and chlorine is weak and liable to break down in sunlight so stabilisers have to be added. A good example is PVC plastic drainpipes which would break up quite quickly without the stabiliser. Most of the other common polymers are sensitive to ultraviolet light to some extent.

However, it cannot be denied that there is a disposal problem. Plastics now account for around 7 per cent by weight of domestic refuse and many people argue that plastic packaging should be **biodegradable** (that is, made so that it can be broken down by bacteria). One way of making a biodegradable plastic is to build sugar links into it, so that the short chains of the original polymer are joined together by sugar molecules to make up the finished chain. Bacteria can then digest the sugar, leaving much smaller chains which will easily degrade in the environment.

In designing biodegradable polymers, we have to be careful that any chemical resulting from the degradation process does not present a greater environmental hazard than the original polymer. The effect of many of these by-products on the water cycle, for example, is hard to predict; and widespread use of biodegradable plastics could lead to environmental problems in 20 to 30 years' time.

So if there are still problems of biodegradability to be solved, how should we dispose of plastics? At present most of our rubbish is either burnt or put onto rubbish tips. Incineration at a high temperature is a good answer; tipping is less desirable as many plastics take a very long time to break down in the soil. **Pyrolysis** has been suggested as a solution: using this method the plastics are burnt under carefully controlled conditions to recover the original monomers which can then be recycled to make further polymers.

Biodegradable polymers already have some important applications in the medical field. The stitches used to sew a wound after an injury or operation are now often biodegradable so there is no need to have them 'out' afterwards. Certain cancer control drugs can also be incorporated into a polymer chain, a pellet of which is then injected under the skin. As the polymer slowly degrades the drug is released. It is this type of slow drug release which some people are suggesting should be used for the chemical castration of sex offenders. There have also been some suggestions that surgical swabs should be biodegradable so that there is no problem if they are left in the body after an operation!

Plastics in the future

With a great deal of research being undertaken, plastics have an exciting future in many areas of our lives. Most present-day polymers are known as **homopolymers** – that is, polymers made from only one monomer. The addition of sugar to make a polymer biodegradable is an example of a **copolymer** (a polymer which contains two different basic monomers). Because the arrangement of the monomers in copolymers can vary (they can go alternatively or in pairs, for example), many different chains, with different properties, can be made from the same monomers.

ABS is another copolymer – a polymer chain made up of about 80 per cent **styrene** monomer and 20 per cent rubber. Virtually all modern telephones, for example, are made from this material. Polystyrene is strong but not very flexible – by adding rubber the material can be made more durable and able to withstand heavy use in a busy office.

Scientists hope that by continuing research into copolymers they will find better substitutes for natural polymers. This could be useful in finding new synthetic fibres for clothing. Present-day synthetics are generally blended with natural fibres (e.g. polyester/cotton) because if used on their own they are not able to 'breathe'. What we really need are fabrics with the feel and performance of cotton or wool but which are also as easy to clean as polyester.

In medicine, polymers may well be used as the base material for artificial organs such as the liver. Indeed early experiments with artificial hearts have already relied on them.

Polymers do not perform well at high temperatures, such as you might find in a car engine, so high-temperature polymers that can withstand temperatures of up to 850 degrees Celsius are now being developed. Maybe one day plastics will really become a substitute for steel . . .

7 HEALTHY EATING
The facts about food

Alan Stewart

> **This chapter investigates:**
> - nutrition
> - the role of nutrients
> - nutritional deficiencies
> - food allergies and intolerance
> - heart/circulatory disease and diet
> - cancer and diet
> - eating heathily
> - diet in the future

What is nutrition?

Nutrition is becoming everybody's business. We all have bodies, we all eat, and we all have our own opinions about food, what we like, or dislike, and what is good for us. Added to this, there is all the advice we get from government committees, food manufacturers and supermarkets. Somehow, amongst all this information and advice, we the consumers make daily decisions about what to eat.

Before this century, eating habits were determined very much by what was available. Many foods could only be obtained for short seasonal periods and local or home-grown produce was an important part of the food supply. After the Second World War, and the end of rationing in the 1950s, we began to turn from eating whatever was available to eating what we liked.

Now that food is in ready abundance, we are all beginning to gain an increased awareness of its importance for our health. We are choosing foods not only because they are pleasant to eat, but also because we believe that they are good for us.

Nutrients The body can be regarded as a machine, which, like any other machine, is made up of component parts working together to perform a number of specific functions. The food we eat contains certain nutrients which are essential for the normal functioning of this machine, and these essential nutrients include **protein**, fats, **carbohydrates**, water, vitamins and minerals. Take one or more of these nutrients away, and to a greater or lesser degree, the body suffers. With extreme deficiency of a nutrient the body can eventually die.

The effects of a lack of calories have been painfully obvious to us all in recent television and newspaper reports on famine-stricken developing countries. However, this is not the problem in the West, except in isolated cases. In this part of the world we are usually dealing with imbalances – too much in the way of calories, fats, refined carbohydrates, alcohol or, on occasions, lack of specific essential vitamins and minerals.

Fats Most of our energy comes from fats and carbohydrates. Fats provide 9 calories per gram, carbohydrates and proteins 4 calories per gram of food. So fats contain more than twice the calories provided by a similar weight of carbohydrate- or protein-based food. In the West, about 40 per cent of our calorie intake comes from fats, usually in the form of dairy products and meat.

Vegetable oils provide only 33 per cent or less of our fat intake, and are mainly found in vegetable oils used in cooking, margarine, nuts and seeds. Smaller quantities are also found in beans, lentils and grains, such as wheat and oats. Vegetable oils, with the exception of palm oil and coconut oil, are mainly **polyunsaturated**. This means that, as well as providing calories, they can also be involved in important biochemical reactions within the body. In fact, two types of polyunsaturated oils (**linoleic acid** and **linolenic acid**, known as **essential fatty acids**) are essential for health.

Saturated fats, which are found mainly in animal products, such as butter and cheese, are simply a source of calories. They remain solid at room temperature, unlike polyunsaturated fats, which always remain liquid at room temperature.

Cholesterol is a particular form of animal fat, which we associate with heart disease. It is not saturated or polyunsaturated, although foods which are high in saturated fats are usually high in cholesterol as well. Two-thirds of the cholesterol in our bloodstream is actually made by our own metabolism, although increased amounts of blood cholesterol are produced if we consume a diet rich in saturated animal fats and low in **fibre**.

Carbohydrates Carbohydrates are organic compounds formed from carbon, hydrogen and oxygen. **Glucose** and **fructose** (fruit sugar) are **simple sugars,** and table sugar, or **sucrose**, is in fact a combination of glucose and fructose. Several different simple sugars may be joined together, and these are called **complex carbohydrates**.

Complex carbohydrates are found in all vegetables and fruits. During the ripening process, the complex, non-sweet-tasting carbohydrates are gradually broken down into simple sweet-tasting carbohydrates such as fructose. A ripe banana, for example, contains much more fructose and less complex carbohydrate than an unripe one. Not all complex carbohydrates can be broken down by the human digestive system, and these 'left-over' carbohydrates are known as fibre. The term 'carbohydrates' therefore includes simple carbohydrates (sugars), complex carbohydrates and complex indigestible carbohydrates which form fibre.

Proteins Proteins are natural polymers – large molecules made up of individual **amino acid** monomers. Different types of proteins are formed by long chains of amino acids, rather like beads on a necklace, which are then twisted and turned into specific shapes.

During digestion, proteins are broken down into their individual amino acids, which are absorbed into the body and then re-made into different types of protein. The proteins found in the human body are very different from those found in other animals or plants. If the process of digestion and re-structuring did not take place, we would begin to take on the cell-structure, and perhaps appearance, of the type of food we ate. Those who ate a lot of chicken might begin to grow feathers, and those who ate a lot of fish might notice a distinct roughening of their skin!

There are many types of amino acids. A normal diet provides about 20 of them, of which eight are essential. Animal products contain all the essential amino acids, but vegetarian proteins may provide only some of them. It is therefore important that vegetarian proteins are mixed (beans with rice, for instance) to balance the intake of amino acids.

Vitamins At the beginning of this century it was discovered that vitamins were essential for animal and human life. Only tiny quantities of these compounds are required in the diet, yet they have an enormous effect on the healthy functioning of the body.

Box 1   Vitamins

Fat-soluble vitamins *What they do*

vitamin A — needed for healthy skin, eyes and resistance to infection
vitamin D — needed for healthy teeth and bones
vitamin E — needed to protect tissue against wear and tear
vitamin K — needed to prevent excessive bleeding or bruising

Water-soluble vitamins

vitamin C — needed to protect tissue against wear and tear, and heal wounds (deficiency leads to easy bruising and bleeding from the gums)

vitamin B complex:
* B1 – thiamin
  B2 – riboflavin
  B3 – nicotinamide
  B5 – pantothenic acid
* B6 – pyridoxine
* folic acid
* B12 – cyanocobalamin

needed for many important functions including the production of energy and the breakdown or formation of proteins, fats and carbohydrates in the body (deficiency affects the brain and nerves – causing changes in emotional state – memory and concentration, the skin, and the blood – producing anaemia)

Deficiencies are rare, except those marked *.

Each vitamin controls several important chemical reactions within the body, and there is probably quite a variation in how much of any one vitamin is needed by an individual.

There are two sorts of vitamins: those that are soluble in fat and the water-soluble vitamins. In times of plenty, there are usually substantial stores of the fat-soluble vitamins. Water-soluble vitamins are stored in much smaller quantities, so deficiency may develop more rapidly – the only exception to this being vitamin B12, which can remain in the liver in substantial amounts. Sudden illness, change in diet, a period of weight loss, or an alcoholic binge, can cause significant B vitamin deficiencies within a few weeks, whereas other nutritional deficiencies are likely to develop over a much longer period of time.

Minerals The essential minerals are of two sorts; the **bulk minerals** which are required in large amounts (several grams or large fractions of a gram per day) and the **trace minerals** which are required in tiny amounts (thousandths or millionths of a gram per day). Bulk minerals are important for the structure of the body and its metabolism.

Trace minerals, together with vitamins, alter the character and speed of the chemical reactions that occur within the cells, and they do this by influencing the activity of **enzymes**. (Enzymes determine which chemical reactions occur inside which cells.)

Nutritional deficiencies

The body runs on the same basis as an accountant's balance sheet. Income must exceed or equal expenditure, otherwise a deficiency will develop, rather like the one experienced by Mr Wilkins Micawber in *David Copperfield*. The ever-optimistic Mr Micawber, whose expenditure exceeded his income, was always waiting for something to turn up. Similarly, the robust human metabolism will hang on doggedly, waiting for a further supply of the missing essential nutrient to materialise.

After food has been eaten, it has to be digested, broken down into its component parts and absorbed. Thus, deficiencies of vitamins and minerals can develop because of poor intake, inadequate

Box 2 Minerals

Bulk minerals	What they do
sodium, potassium, chlorine, phosphorus	These minerals form most of the chemical structure of the body, and are all closely involved in many areas of metabolism.
calcium, magnesium	Together with phosphorus, these minerals are essential for healthy bones, teeth, muscles and nerves.

Trace minerals	
iron	essential for blood formation, muscle and brain function
iodine	needed for production of hormones by the thyroid gland (these hormones control the speed of the body's metabolism)
zinc	needed for tissue repair, growth and resistance to infection (high concentrations in the eyes, muscles and ovaries)
chromium	helps control blood sugar level
selenium	together with vitamin E, helps to protect tissues against wear and tear
copper	together with many vitamins and other minerals, influences tissue repair and the blood
cobalt	part of vitamin B12 and needed in the formation of red blood cells

Note: Iron deficiency is one of the most common deficiencies in both western and developing countries.

digestion or increased utilisation or losses, perhaps due to diseases of the liver, pancreas, small bowel or kidney, alcoholism or drug use. Furthermore, the diet may also contain **anti-nutrients** which can obstruct the absorption of many nutrients (*see Box 3*). The effects vary considerably from one individual to another but in all these situations, doctors can correct the deficiency by increasing the intake, correcting problems with absorption or treating the underlying disease.

What are the symptoms of deficiencies? These can include mental symptoms, such as depression, anxiety or lethargy, and some people will simply feel 'off colour' long before physical signs develop. Some common physical features of vitamin and mineral deficiencies are shown in *Figure 1*.

Box 3 The social anti-nutrients

Anti-nutrient	Associated deficiencies
smoking	vitamin C and vitamin B6
tea or coffee	iron and zinc
alcohol	all nutrients, especially vitamin B complex, magnesium and zinc
bran	calcium, magnesium, iron and zinc
drugs:	
diuretics	potassium and magnesium
the birth-control Pill	vitamin B6, B complex vitamins and possibly zinc
anti-epileptic drugs	vitamin D and folic acid

Nutritional supplements

The majority of us who are well, and eat a healthy diet, do not require **nutritional supplements**. There may, however, be exceptions to this, such as elderly people, who are sometimes found to have mild deficiencies, which could increase the likelihood of illness in the future. Such individuals, as well as those on weight-reducing diets, or those receiving medical care, should take an appropriate multivitamin and multimineral supplement, preferably under a doctor's direction.

Food allergy and intolerance

Allergic reactions have become a very popular topic over the last 10 years. Most people think of allergies in terms of conditions like hay fever, where sensitivity to grass or other pollens produces eye irritation, nasal irritation and sneezing.

However, the idea that people can be allergic to foods is relatively new. In some cases, eating nuts, strawberries or shellfish can result in swelling of the lips and tongue within seconds or minutes, and in these situations it is easy to identify the food which has caused the condition. But some reactions may only take place after several hours or even one or two days, and produce symptoms in parts of the body which are distant from the mouth. Conditions such as asthma, rheumatoid arthritis and widespread eczema in both children and adults can all be caused by food allergies – as well as other factors.

HEALTHY EATING

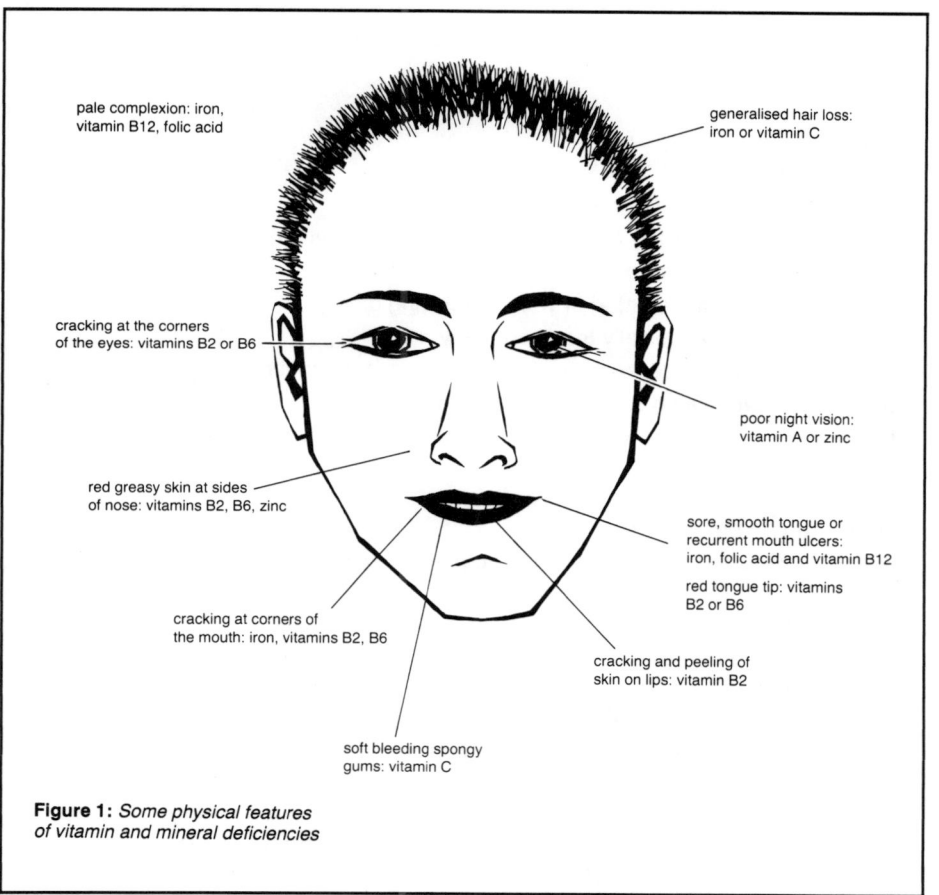

Figure 1: *Some physical features of vitamin and mineral deficiencies*

It is thought that some of the food, instead of being fully digested and broken down into its basic nutritional components, is absorbed intact into the bloodstream. The body reacts to this alien substance as if it were a foreign germ, and produces antibodies and white cells which interact with the food. We do not know why this reaction occurs in the first place, nor what part of the body it takes place in.

Many other people, however, have what is known as **food intolerance** – that is, they are unable to tolerate certain foods because of an adverse reaction which does not necessarily involve

> **Box 4 Those at risk of developing nutritional deficiencies**
>
> *Multiple deficiencies*
>
> premature infants
> children during growth spurts
> pregnant or breast-feeding women
> elderly people, especially those living alone or in institutions
> anyone losing weight due to illness or poor appetite
> those on long-term restricted or weight-reducing diets
> the chronically ill, especially those on many drugs
> alcoholics or regular consumers of alcohol in large amounts
> patients with advanced cancer
> those with any serious medical illness, especially liver or gastro-intestinal disease
>
> *Specific deficiencies*
>
> vegetarians (*iron and zinc*)
> vegans – vegetarians not consuming eggs or dairy products (*iron, zinc and vitamin B12*)
> smokers (*vitamin C*)
> dark-skinned, urban-dwellers, especially when pregnant or breast-feeding (*vitamin D*)
> housebound or institutionalised elderly people (*vitamin D*)
> psychiatric in- or out-patients (*vitamin B complex and vitamin C*)

antibodies or white cells. Such adverse reactions are often found in people who suffer from migraine, irritable bowel syndrome (a condition characterised by intermittent abdominal bloating and discomfort, with either diarrhoea or constipation) and children who are hyperactive.

Why healthy, nutritious foods should produce adverse reactions in otherwise healthy individuals, is uncertain. We have yet to identify the factors which have led to the development of allergies.

The foods which most commonly produce food allergy or intolerance are cow's milk, cheese, wheat and other grains, corn, food additives, eggs, fish, alcoholic drinks, tea, coffee, chocolate, oranges, yeast and foods made from or containing yeast extract.

HEALTHY EATING

Heart/circulatory disease and diet

Most people are aware that there are links between diet and diseases of the heart and blood vessels (**the cardiovascular system**). To give a few examples, high blood pressure (or **hypertension**) and **coronary artery disease** (which lead to strokes and heart attacks respectively) are both influenced by the type of food we eat.

High blood pressure

Blood pressure usually rises in the West as people get older, and high blood pressure is now very common in developed countries. When the doctor takes your blood pressure, he records two figures: the top one relates to the highest pressure generated when the heart contracts, and the lower one represents the resting pressure found in the blood vessels when the heart relaxes. Normal blood pressure is below 140/90 (millimetres of mercury) for a young to middle-aged adult, and levels above this are associated with an increased risk of strokes and heart disease.

In many cases, the cause of hypertension is uncertain, but diet is now known to be a substantial factor. Though many powerful drugs have been developed to control high blood pressure, recent research has shown that up to 40 per cent of individuals with hypertension may be able to control their condition through losing weight, avoiding alcohol and reducing the amount of sodium salt in their diet – both in the food and at the table.

Coronary artery disease

The coronary arteries are two specialised arteries that supply the heart muscle with blood and oxygen, as it continually contracts and relaxes to pump blood around the body. If these arteries get 'furred up', the blood supply is reduced, and a severe, cramp-like pain is felt in the chest. If this continues, part of the heart muscle will die.

There are two main factors which lead to this type of coronary artery disease. The first is the build-up of cholesterol in the wall of the coronary artery. This reduces the size of the artery, and therefore the flow of blood to the heart muscle. Secondly the blood sticks to the cholesterol deposit, and a blood clot (or thrombus) forms, completely obstructing the coronary artery. Exercise improves heart function, lowers blood cholesterol, and helps to reduce obesity. In contrast, alcohol has an adverse effect upon blood fats, and excessive consumption (more than 3 units of alcohol – half a bottle of wine or $1^{1}/_{2}$ pints of beer per day) is associated with increased coronary heart disease risk.

We still cannot be certain whether it is possible to reduce the risk of coronary heart disease by changing our diet, but the rates of coronary heart disease have indeed been falling in North America and Finland. At times this fall has coincided with public health measures to reduce coronary heart disease risk by improving the nation's diet, but the picture is not entirely clear-cut or convincing.

Stroke

Strokes are caused by a haemorrhage or clot in the vessels supplying blood to the brain. The risk of stroke has been falling in this country since the Second World War and this is almost certainly due to better blood pressure control and perhaps improved vitamin C intake from fruit and vegetables. Vitamin C seems to be helpful in strengthening blood vessels, and thus reduces the risk of haemorrhage.

Cancer and diet

Cancers of various types, together with cardiovascular disease, are the most common causes of death in Western society. A cancer is a group of cells which has broken away from the normal growth control mechanisms. These new cells can grow and spread, destroying healthy tissue – often with disastrous consequences.

Environmental factors play a large part in the type of cancers found in society. Smoking, in particular, tends to lead to cancer of the lung, but also increases the risk of cancer of the mouth, throat and bladder. Alcohol, radiation and occupational factors influence the risk of certain types of cancer, and in roughly a third of cases, the level of risk is affected by diet. In this respect, an excess of fats, particularly saturated animal fats, may increase the risk of cancer of the breast, and adequate amounts of certain nutrients (notably vitamin A, selenium and vitamin E) may

HEALTHY EATING

help to prevent the formation of certain types of cancer. Furthermore, a high intake of fibre may reduce the risk of bowel cancer.

Overall, the potentially cancer-preventing nutrients are to be found in significant quantities in fresh green leafy vegetables, which should be a regular part of everybody's diet.

Many of us have already made changes to our diets over the last 10 years, as shown in the bar charts comparing eating habits in 1976 and 1984. All income groups seem to have got the message about wholemeal bread and sugar. We have increased our intake of cereal fibre and reduced our consumption of sugar. Overall, most people have cut down on fats by about 10 per cent, so we are well on the way to achieving the dietary goals that many experts would recommend. The only exceptions are fresh fruit and vegetables. Income groups A, B and C (the more prosperous members of society) eat between 5 and 10 per cent more fruit and vegetables. However the poorer members of the community (income group D) have reduced their consumption by 10 to 15 per cent.

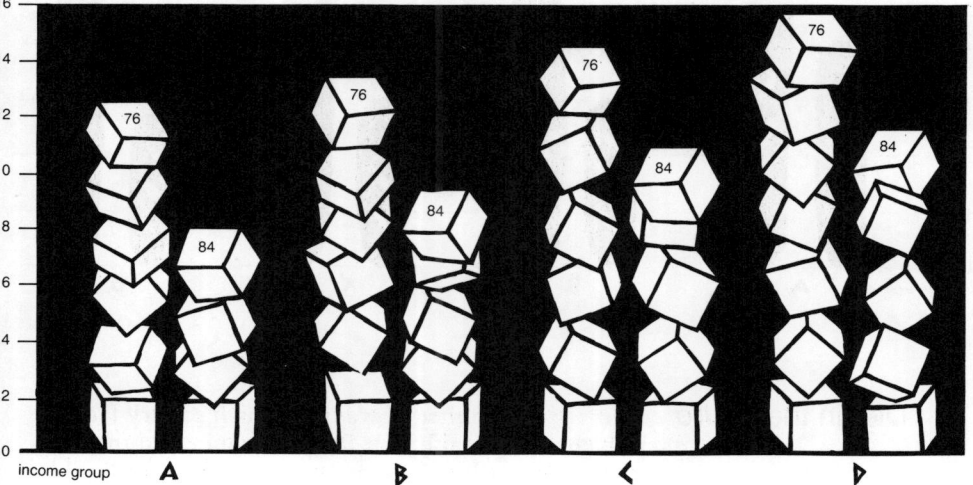

Figure 2: *Sugar consumption in 1976 and 1984*

107

INSIDE SCIENCE

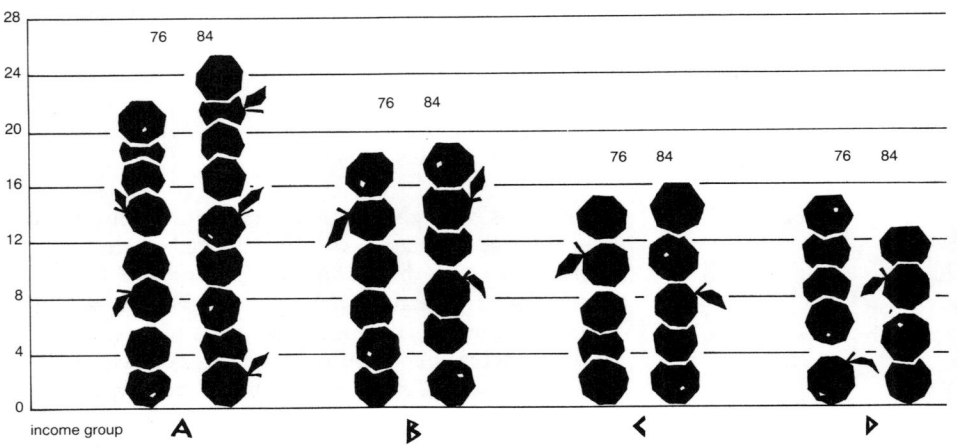

Figure 3: *Fresh fruit consumption in 1976 and 1984*

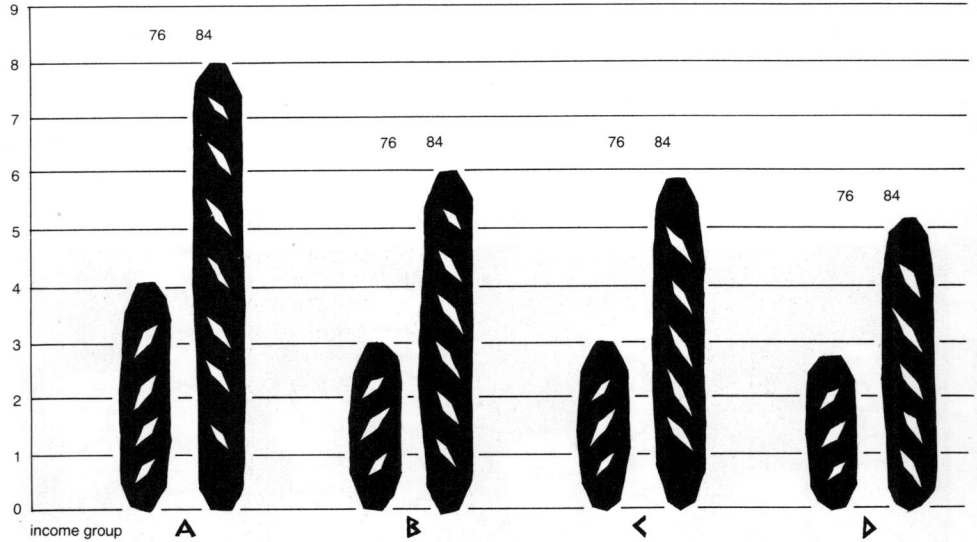

Figure 4: *Brown and wholemeal bread consumption in 1976 and 1984*

Diet in the future

There are a number of ways in which dietary factors influence our health, our life expectancy and the quality of our life. In some ways the Western world

HEALTHY EATING

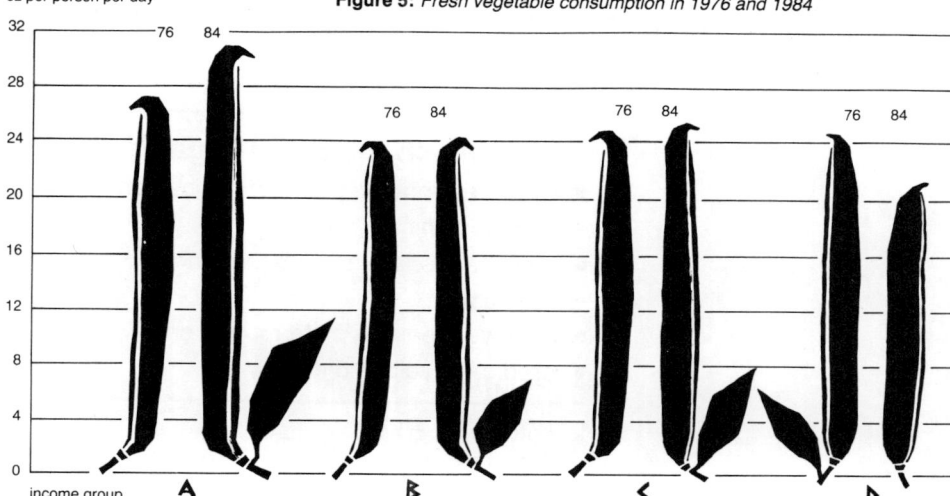

Figure 5: Fresh vegetable consumption in 1976 and 1984

is nutritionally better-off than ever before. We have abundant access to cheap food of excellent quality, unrivalled transport and storage systems, and the power to communicate through education, the media and advertising about the merits or otherwise of different types of food. It is perhaps this latter point, education, that will finally determine how our diets change over the next 20 or 30 years.

Box 5   Eating healthily

Here are some simple guidelines for healthy eating, based on 'average' consumption:

- If you are overweight, try to reduce to your ideal weight.
- Limit alcohol intake to 1–2 units of alcohol per day for men, and ½–1 unit per day for women. (1 unit of alcohol is equivalent to one glass of wine, one serving of spirit or vermouth, or half a pint of standard strength beer or lager.)
- On average, you should aim to cut down your fat intake (particularly animal fat) to 75 per cent of its current level.
- Sugar (or sucrose) consumption should be reduced to about 75 per cent of your present intake.
- Fibre (from fruit, vegetables and wholegrain cereals) should be increased by about 50 per cent.

8 THE CHANGING EARTH
Eruptions and earthquakes

Hazel Rymer

> **This chapter investigates:**
> - where earthquakes and volcanoes occur
> - plate tectonics
> - what happens at plate boundaries
> - hot spots
> - how we measure plate movement
> - predicting volcanic activity

Volcanoes and earthquakes have fascinated civilisations through the ages. In historic times their effects were often put down to fiery and angry gods. And the effects can certainly be spectacular. Pompeii was buried in minutes by a layer of volcanic ash several metres thick from Mount Vesuvius in AD 79. In 1883 the explosive eruption of Krakatoa covered 4 million square kilometres of land with ash, led to coloured sunsets around the world for several years and produced a tidal wave which killed 36 000 people on nearby islands. These dramatic events are not just confined to the past: since the beginning of this century earthquakes have killed over 1 million people, more than half a million dying in a single earthquake in 1976 – the Tangshan earthquake in China.

The damage caused by these huge events can be long-term as well as immediate. It is now thought that volcanic ash and gases thrown up into the atmosphere during the biggest eruptions may significantly cool the whole planet. Prehistoric eruptions are thought to have had devastating effects on the Earth, causing global temperatures to fall by tens of degrees for months or even years. Indeed some geologists suggest that dinosaurs may have become extinct as a result of a **volcanic winter** during which the Earth was shaded from sunlight for many months.

Box 1 A brief history of the Earth

Our sun, and its planets were created about 4600 million years ago. The early solar system consisted of a swirling mass of hot gases, some of which condensed, and formed dust and finally rock fragments. These fragments collided together, producing a nucleus to which more material was then attracted by the force of gravity. As it formed, the centre of the planet became hotter. Gradually the Earth took on its present form, shown in *Figure 1*, an iron-rich core covered by a mantle of solid rock material. Roughly 150 kilometres below the surface the mantle contains about 5 per cent molten rock and the rest is solid. The molten rock is lighter than the solid, so it rises to the surface where it cools to form the solid crust that we live on.

In those early days, volcanic activity was more widespread and vigorous than it is today, and much of the matter emitted by volcanoes consisted of gases – generally a mixture of about 80 per cent steam, 10 per cent carbon dioxide and 4 per cent nitrogen, together with a variety of other gases. The steam condensed to form the oceans with some dissolved carbon dioxide. Plant **photosynthesis** has converted carbon dioxide into oxygen and this, together with nitrogen and other gases, has formed the life-supporting atmosphere of our planet.

Figure 1, *below: Structure and composition of the Earth's interior*

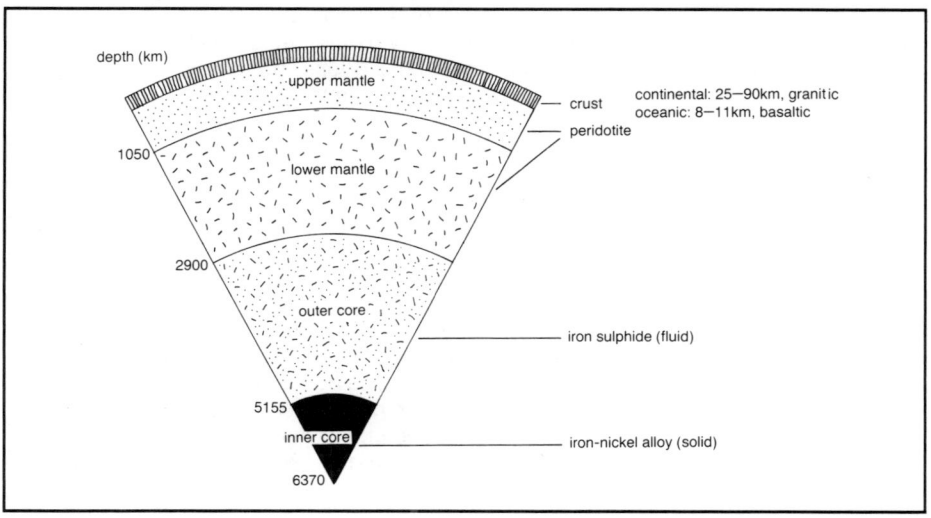

There is also a good side to the effects of volcanoes which we should not forget. Volcanic ash quickly weathers to form highly fertile soil which is excellent for growing a wide range of crops. For this reason, in many countries, the most highly prized land is on volcanic slopes. Another benefit is that in many volcanic areas molten rock heats underground water to form steam which, when it is near the surface (as in Tuscany, Iceland and New Zealand), can be used to provide domestic heating or drive turbines to generate electricity.

Box 2   Geothermal power

In volcanic areas, underground water may be at a temperature of more than 200 degrees Celsius within 1 kilometre of the surface. Wells can be drilled, in the same way as oil wells, to extract the steam which is then used to drive turbines to generate electricity. This is one of the world's fastest-growing sources of energy. Since the water is heated by enormous reservoirs of cooling rock, several kilometres across, geothermal steam can be considered as a renewable source of energy. Even in non-volcanic areas, underground water may be heated by natural radioactivity, although it is necessary to drill deeper (to around 6 kilometres) before the water is at a temperature of 200 degrees Celsius. In Britain there are no natural reservoirs of volcanic steam, but the granites in Cornwall are suitable for this type of 'hot dry rock' **geothermal power** production in which the heat of the rock is used to produce steam from water pumped to deep levels.

Heat energy can also be extracted from recent lava flows, but it does not last as long. In 1973 the island of Heimay, just south of Iceland, was expanded by an eruption of basaltic lava and ash. The lava flow is some 30 metres thick, and still contains so much heat that when cold water is pumped in at one end of the cooling lava, hot water comes out at the other. This hot water is then used to heat all the homes on the island.

Where do earthquakes and volcanoes occur?

Most of the world's most destructive earthquakes and volcanoes occur in well-defined narrow zones. For example, about 60 per cent of large earthquakes are associated with active volcanoes around the borders of the Pacific Ocean. The most devastating historical earthquakes – the remaining 40 per cent – occur in a strip stretching from the Mediterranean countries, through the Middle East and the Himalayas, to Indonesia and southern China. There are also distinct zones within the oceans where this sort of activity occurs; these zones are known as **oceanic ridges** and generally consist of underwater mountains some 1 to 2 kilometres high. The tops of these underwater mountains sometimes form islands, like Iceland.

Zones of volcanic and earthquake activity are important because they mark the boundaries or joins, between different sections of the Earth's rigid outer skin. This outer skin is known as the **lithosphere**. The lithosphere is very stiff compared with the hotter regions below and on average it is about 100 kilometres thick below the oceans and 200 kilometres thick beneath the continents. It is made up of two layers: the **crust** and the **upper mantle**, as shown in *Figure 1*.

The different sections of this rigid lithosphere are known as 'plates', most of which have both continental and oceanic parts. The oceanic part of a plate is denser than the continental part so it presses down harder on the mantle below, forming the ocean basins. There are nine major and many smaller plates covering the Earth's surface, and these are shown in *Figure 2*. The study of the movement of plates is known as **plate tectonics**.

Plate tectonics

The theory of plate tectonics is only about 25 years old and was developed after the invention of sophisticated measuring instruments which allowed us to measure the relative movement of the different parts of the Earth's surface. With this theory we have been able to understand the history of our planet more clearly.

Plate tectonics theory tells us that the movement

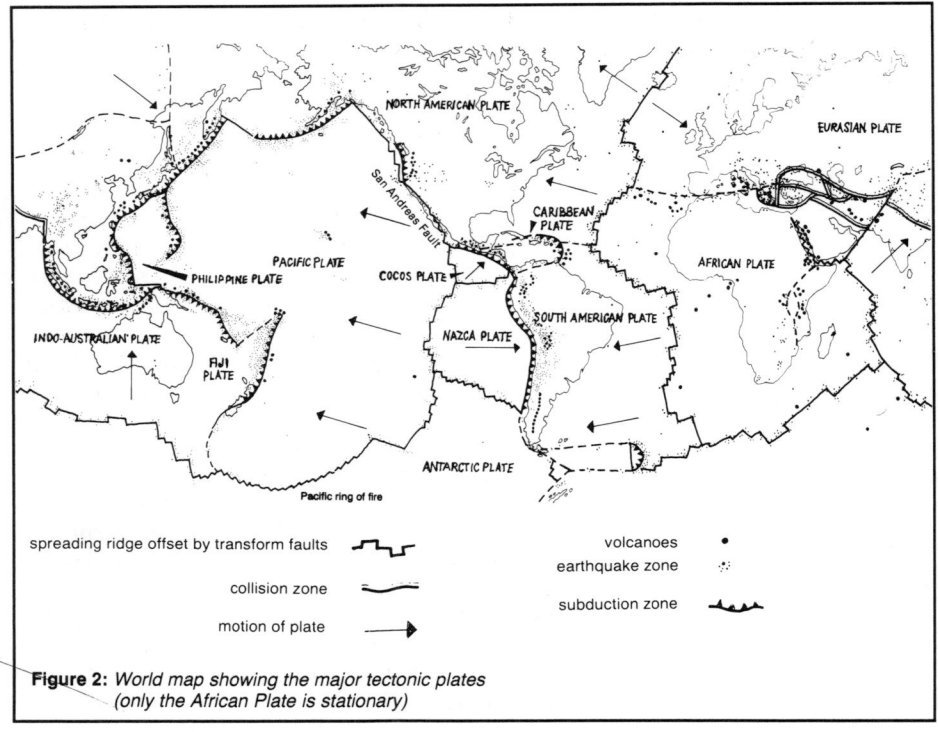

Figure 2: *World map showing the major tectonic plates (only the African Plate is stationary)*

of the plates is dictated by the convection currents in the hot, deformable rock, deep within the mantle. These currents are similar to those which can be seen in a pan of near-boiling soup, caused by the high-temperature soup moving up to the surface from the bottom of the pan. Of course the plates have enormous mass so they only move by about a centimetre per year, but even this degree of movement can have dramatic effects.

It is at the plate boundaries (where the plates move apart, collide or push past one another) that earthquakes occur and volcanoes are formed. One example of this can be seen as the south part of the American Plate moves towards the Nazca Plate. There, the friction between the plates causes earthquakes and the heat produced by the friction causes a partial melting of buried rock. As the rock begins to melt it becomes lighter and rises towards the surface, causing a volcanic eruption.

What happens at a plate boundary?

When two plates come together, the heavier, denser one is pushed down below the lighter one. This is the process of **subduction** at a **destructive plate boundary** shown in *Figure 3*. Because the cold, rigid plate being pushed down into the hot mantle is far denser than the surrounding hotter material, it continues to sink downwards, usually at an angle between 45 and 70 degrees below the horizontal. As the heavier plate moves down, friction and other effects such as thermal expansion cause earthquakes as deep as 700 kilometres below the surface. During subduction a trench is formed on the surface at the junction between the plates. Andesite, a light crystalline rock, comes to the surface, erupting through volcanoes which are found on the edge of the lighter, overlying plate.

The simplest instance of a destructive plate boundary is an **island arc**, formed by the collision of two oceanic plates. The Caribbean islands are one example, formed by the Atlantic Plate (which is moving west at about 1 to 2 centimetres per year) being subducted under the eastern edge of the Caribbean Plate. Most of the volcanoes in this arc are too small to break the ocean surface, but some have formed islands large enough to support human populations. Closer to home, a once-active island arc can be seen in the 450- to 500-million-year-old volcanic rocks of the English Lake District and

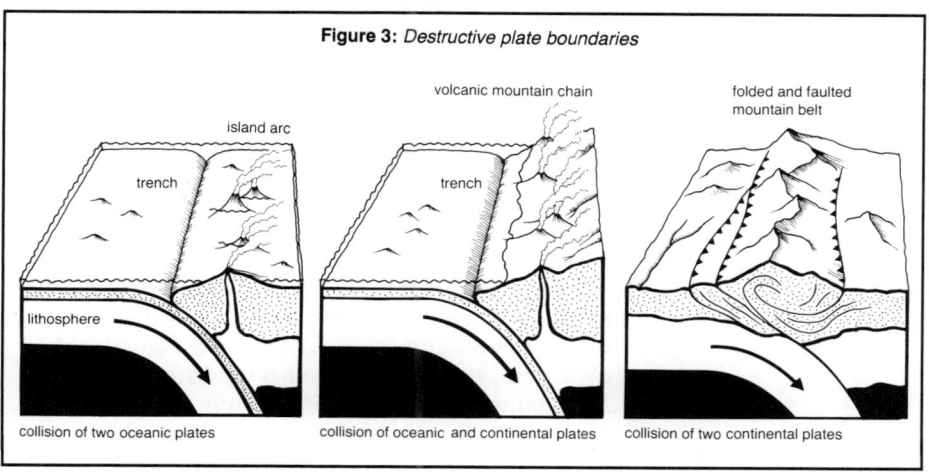

Figure 3: *Destructive plate boundaries*

similar rocks which are to be found in North Wales.

When the continental parts of two plates are forced together, neither can be subducted below the other because they are both formed from the same material. Instead there is a **continental collision**. As the plates press together, the edges buckle and the huge pressure causes dramatic changes in the rocks, both buried and on the surface. This is accompanied by shallow earthquakes. The Himalayas, for instance, are the result of India and Asia colliding in this way.

As well as being pressed together, plates can move apart, and the places where this happens are known as **constructive plate boundaries**. Here, as the crust opens, there is a partial melting in the upper mantle to produce basalt (a dense black rock) which rises to fill the gap in the crust, creating new oceanic crust. As the plates move apart, a line of basaltic volcanoes, known as an oceanic ridge, forms to fill the gap and so a new ocean grows in a process known as **ocean floor spreading**. Molten basalt is thinner and more runny than the andesite found at destructive plate boundaries so it covers the surface with lava flows rather than forming the steep-sided volcanoes characteristic of andesite. The process, where the formation of new oceanic crust pushes continents apart, is known as **continental drift**. Over millions of years continents can move enormous distances.

The most striking example of continental drift can be seen in the South Atlantic Ocean which began to open up about 120 million years ago. Before that, South America and South Africa were joined together, and this can be seen from the way their Atlantic coastlines look as if they once fitted together like pieces from a gigantic jigsaw. *Figure 4* shows that the resulting oceanic ridge lies symmetrically between the two continents.

Although it is not as obvious, Britain has also moved. When the forests making up our coal reserves were alive, it was close to the equator, but over millions of years continental drift has moved Britain north to its present position.

THE CHANGING EARTH

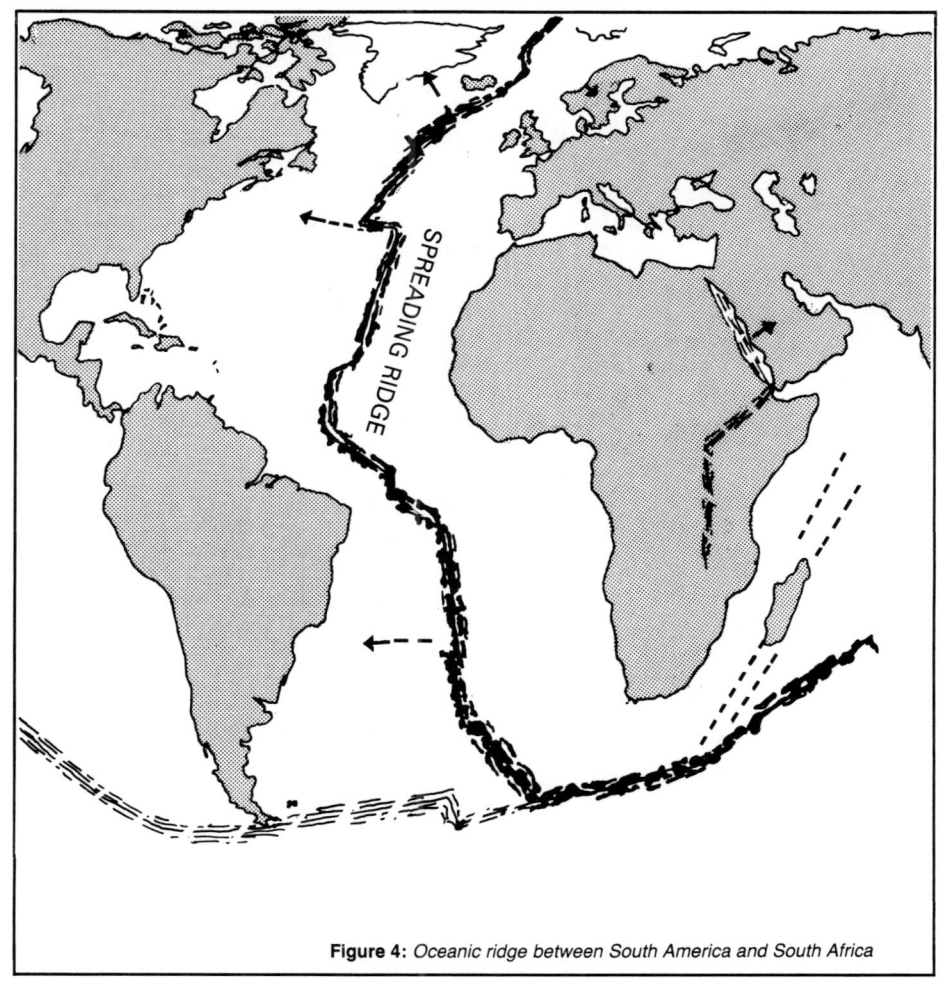

Figure 4: *Oceanic ridge between South America and South Africa*

Hot spots So far we have discussed two types of volcanoes: the andesite ones at destructive plate edges and the basaltic ones at constructive plate edges. There is a third type: the volcanoes that occur within plates at what are called **hot spots**, usually well away from the plate edge. These occur because some parts of the Earth's mantle are hotter than normal and their heat burns through even the thickest continental lithosphere to produce basaltic volcanoes.

The Hawaiian volcanic chain is a string of islands and submerged volcanoes in the middle of the Pacific Plate. The north-westerly movement of this plate over a stationary hot spot has produced a chain of volcanoes that are younger towards the south-eastern end of the line. The older, north-western part of the chain is below sea level but the active southern island (Hawaii) is above the hot spot. The next island is already forming on the south-east coast of Hawaii. Loila volcano is now 2 kilometres high but has another kilometre to go before it breaks the Pacific Ocean surface – this is expected to happen within a few tens of thousands of years.

If a hot spot is burning through the continental lithosphere it is possible for the enormous heat to cause the region to rise up and crack open. Basalt then erupts and forms a narrow zone of new oceanic lithosphere. If the continental masses are then able to move apart a new ocean is formed by sea floor spreading, and this is how the Red Sea was formed, as shown in *Figure 5*.

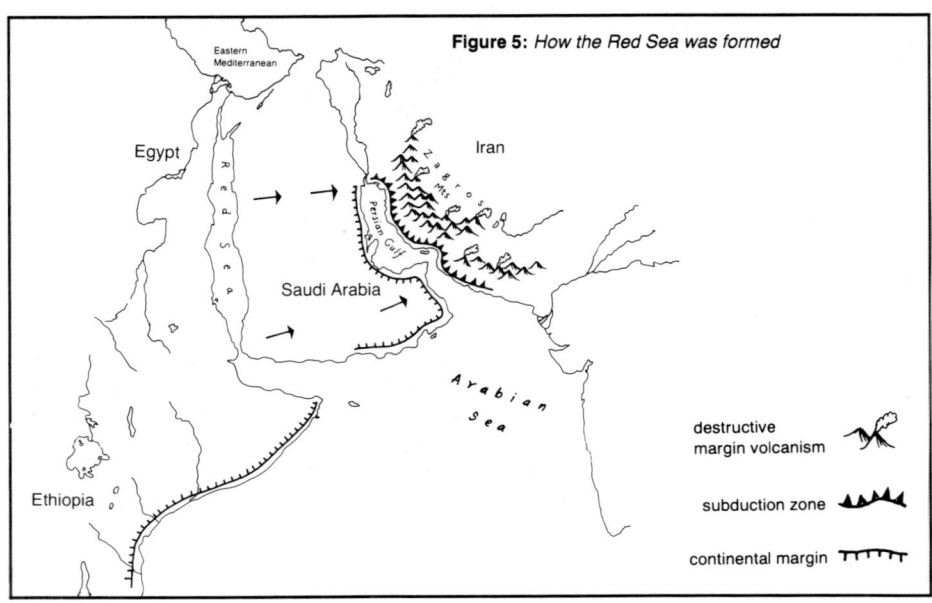

Figure 5: *How the Red Sea was formed*

Box 3 The British example

The ancient Iapetus Ocean, which once separated the American and European continents, closed about 450 million years ago. At that time the Iapetus Ocean was completely subducted beneath the American and European Plates which became a single plate and moved together for about 400 million years. It was at this time that the mountains of North Wales, the Lake District and the Peak District were formed.

About 50 million years ago, east-west tensions across the European-American continent, caused by plate movements in other parts of the world, started to split the continents, leading to the formation of the Atlantic Ocean. The Hebridean islands of Skye and Mull, the Mountains of Mourne in Ireland and Lundy Island in the Bristol Channel are all the eroded remains of volcanoes located along these cracks, each of which was formed by hot spot activity. Activity at these volcanoes was soon replaced by ocean crust formation and sea floor spreading above a line of hot spots further west, as the two continents split apart. And the Atlantic has been widening ever since. With this new split, part of the original American continent was left behind to form part of northern Scotland, above Edinburgh and Glasgow.

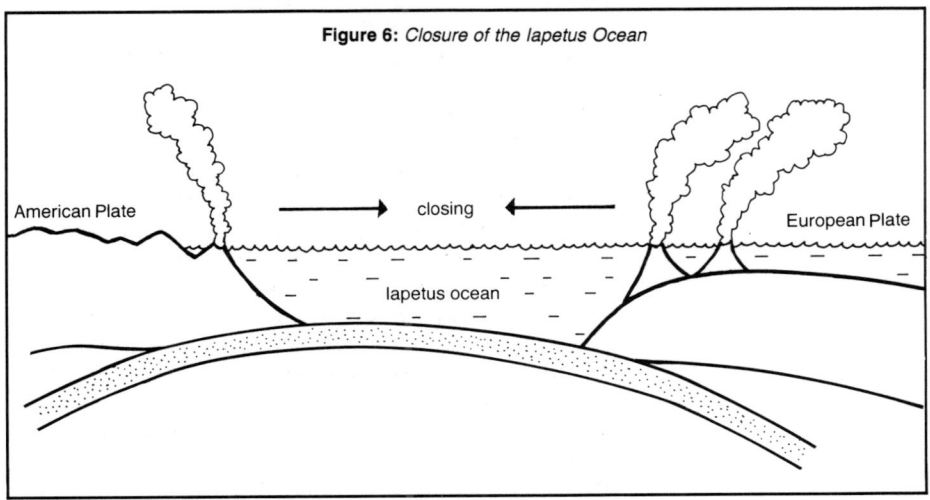

Figure 6: *Closure of the Iapetus Ocean*

How do we measure plate movement?

Earthquakes can be detected by anything sensitive to ground movements of less than a thousandth of a millimetre. A **seismometer** works on the principle that a heavy weight will remain stationary during an earthquake but the rest of the instrument, in direct contact with the ground, will vibrate. The vibration can then be recorded as an ink trace on paper on a rotating drum. A single seismometer will record the time at which the shock wave arrives at a location and its magnitude, which is described on the Richter scale. By using several seismometers spread out over several kilometres it is possible to find the location and depth of the movement (the **hypocentre**) because the wave will arrive at each instrument at a slightly different time.

The movement of the mixture of solid and molten rock in the mantle is difficult to predict accurately and this makes it hard to forecast earthquakes (in the same way as weather forecasters have difficulty in predicting air movement). However, strain gauges are used to measure the build-up of energy in places like the San Andreas Fault and this information can give geophysicists some idea of possible future movement.

The movement caused by sea floor spreading is much more difficult to measure directly because it happens intermittently rather than at a constant rate. However we can date rocks to within a few million years by studying radioactive material trapped within rocks. By taking samples from different locations around a ridge we can work out the rough rate of movement.

A better way of measuring the rate of spread is to look at the pattern of magnetism in the Earth's rocks. When molten lava cools, it becomes magnetised in the direction of the Earth's prevailing magnetic field. As the Earth's magnetic poles reverse about once every million years, the magnetic field direction reverses. Thus we can fly across the oceans with special instruments and plot this changing pattern of magnetism frozen within the volcanic rocks of the ocean floor. This method has shown us that the North Atlantic is opening out at about 1 centimetre per year while the south-east Pacific is opening much more rapidly at about 9 centimetres per year.

Prediction of volcanic activity

Volcanic activity can be measured in many different ways but in order to make accurate predictions a comprehensive monitoring programme is needed. However, this is expensive, so extensive monitoring is normally only used in areas of high risk and potential hazard.

Small groups of seismometers can detect volcanic tremor, a low-frequency vibration felt at active volcanoes when molten rock is near the surface. Measurements of electrical properties, such as small voltages within the ground and the resistance to electric current in rocks, also help to monitor the amount of ground water and the quantity of dissolved metals in it. Molten rock close to the surface will reduce the local magnetic field strength slightly and the ground generally tilts before a volcanic eruption. More recently, satellites have been used to map variations in thermal radiation. All these changes can be monitored in order to plot the rise of molten rock below the surface, which sometimes allows us to predict the time of eruption.

Monitoring volcanoes

Various international agencies are combining to plan a decade of hazard reduction in the 1990s because so few of the world's volcanoes can be properly monitored. In effect this will involve putting more resources into developing multi-component monitoring systems which can be used to assess, rather than reduce, the level of hazard posed by some of the world's potentially most dangerous volcanoes. These are on destructive plate boundaries or close to large towns or cities.

The effects of plate tectonics – the external signs of our planet's dynamic interior – are rarely felt by most people, yet for some occasional and random minorities they are devastating. We can only hope that in the future a better scientific understanding of our changing Earth will reduce the number of people who suffer from the effects of earthquakes and volcanoes.

9 BUILDING ROADS
The way ahead

Dennis Gilbert

> **This chapter investigates:**
> - The first roads
> - Highway planning and design
> - Traffic forecasts
> - Route location
> - Design standards
> - Structure of roads
> - Road building materials
> - Maintenance
> - Safety fences
> - Road building in the future

The first roads

Road building (or highway engineering) is as old as history itself, and had achieved a certain sophistication even in ancient civilisation. Strabo recorded that Babylon had paved roads over 4000 years ago and we have details of the construction of their royal processional route built in about 620 BC.

Another impressive example is the stone-faced sloping causeway built by the Pharaoh Cheops in about 2700 BC in order to move the massive limestone blocks used on the Great Pyramid at Giza. Possibly the oldest road in Europe is the dual carriageway (built in about 2000 BC) connecting Cortina in the south-east of Crete with Knossos in the north of the island.

In Britain, although the ridgeways and trackways came into being in about 2800 BC, no formal roads were constructed until Roman times. The Romans are generally acknowledged as the first real road builders and these early roads were noted for their directness: they were usually constructed as a series of straight sections connecting individual military stations. Also, they were generally built on embankments to give a commanding view of the surrounding countryside, giving rise to the term 'highway'.

Transportation in all its forms still presents the modern civil engineer with a wide range of problems. Road engineering is very different from many of the other modern sciences: we measure the life of a road in tens or even hundreds of years, whereas we measure the life of most new products in months. Also, in this electronic age, when microminiaturisation packs an entire computer into a matchbox, the highway engineer still calculates his quantities of materials in millions of cubic metres or tonnes and his labour force in tens of thousands of people.

Highway planning and design

The planning of modern roads has to take into account the close relationship between transport and the way we use land, as well as an estimate of the movement of people and their goods in the future.

Eighteenth-century road construction consisted of heaping more soil on top of the existing mud and hoping that traffic would compact it into a hard surface! Fortunately a number of surveyors and engineers appeared on the scene and applied scientific methods to the problem. These included Pierre Trésaguet in France – often referred to as the father of modern highway engineering – John Metcalf (Blind Jack of Knaresborough), Thomas Telford and John Loudon MacAdam.

During most of the nineteenth century the railway was undoubtedly the prime mover of people and goods over medium and long distances. However, railways did little to stimulate suburban traffic; this was largely carried by horse buses and trams. Even the advent of the motor car at about the turn of the century had little effect on mass movement for some time – restricted, as it was, to the wealthy. This has changed since the Second World War, and now motor vehicles are widely used, both for personal travel and for the movement of goods.

These broad changes in the pattern of movement over the past 200 years provide the historical backdrop to the problem that highway engineers continually face – planning roads to accommodate the traffic. These changing patterns also show that highway engineering has its own internal dynamics, in

INSIDE SCIENCE

Figure 1, *above: The M25*

that the nature of the highway design problems to be solved are continuously changing. For the modern highway engineer the major factors in planning and design decisions are:

- forecasting the future amount of travel on the road
- choosing an appropriate route
- selecting appropriate design standards.

Traffic forecasts

In practice, the problems of route location and forecasting the demand for road space are inextricably linked. Forecasts of future traffic used to be based simply on past trends but this method is unreliable. Often, although not inevitably, the forecasts underestimated the amount of travel. And as the design of the road (particularly the thickness of its base) was based on these forecasts, an underestimate meant a shorter economic life and costly road repairs. Nowadays, we mathematically forecast the future amount of travel on a particular road, using such factors as household income, vehicle ownership level, land use activity, type of transport, travel times and travel costs.

Traffic forecasts are sometimes spectacularly wrong. Also, environmental and other constraints can mean that roads have to be built to lower standards than those dictated by the forecasts. Errors

> **Box 1   Traffic demand**
>
> There are two types of current traffic and three types of future traffic that have to be taken into account when calculating traffic demand.
>
> *Current traffic*
> This is of two types: existing traffic on the proposed new route or widening; and traffic which will divert from other roads to the new route.
>
> *Normal traffic growth*
> This is the increase in traffic volume over time due to rising numbers and use of motor vehicles. Standard estimates exist of such traffic growth.
>
> *Generated traffic*
> Within the first few years, particularly close to urban areas, new traffic appears because of the new highway. This might occur in one of three ways. First, there are entirely new trips because of improved accessibility. Secondly, traffic is created by changes in the type of transport used as a result of the new road – for example, people driving instead of using the railways. Finally, there are the people who previously made trips to different destinations but now decide to change because of the convenience of the new route. Most generated traffic occurs within a relatively short period of time after the road is opened – say one or two years.
>
> *Development traffic*
> With the opening of a new highway, increasing development takes place on nearby land, and this leads to more traffic. This is a relatively long-term component of traffic growth.

can occur in both directions. The M50 in Gloucestershire carries far less traffic than it is designed for, whereas parts of the London orbital motorway, the M25, were close to their practical traffic capacity almost immediately after it opened. Stretches of the M5 in the West Country are sometimes very congested at weekends in the holiday season, even though its average daily flow is still somewhat less than its practical capacity.

We now use complex mathematical models to

attempt to anticipate all these changes. With these models, we try to forecast the traffic flows on the new road for each year it is open, and also the traffic on those roads which the new road may replace. Because of its inherent uncertainties, this modelling process has sometimes been referred to as 'institutionalised crystal ball gazing'.

The forecasts are made over different periods of time, depending on whether we are estimating the **traffic capacity** (15 years), **economic life** (30 years) or the **technical life** (20 to 40 years) of the road. Maintenance, including resurfacing, may take place on a number of occasions during the technical life of a road in order to extend its useful life.

We use this information for three main purposes: firstly, to estimate the benefits of building the road, including time and accident savings; secondly, in designing the road – that is, working out the number of lanes required, and the most appropriate form and layout of junctions. Finally, along with other data, the design traffic volume helps us to decide on the most suitable construction in terms of materials, thickness of base layers and surfacings.

Route location

We have to take into account a number of often conflicting criteria when choosing a route for a new road.

Figure 2 compares the level of a proposed highway with the existing ground level. These days we try to build roads with relatively modest gradients even though the ground level can vary quite significantly,

Figure 2, *below: Balancing cut and fill in road building (A shows original ground level and B shows level of proposed road)*

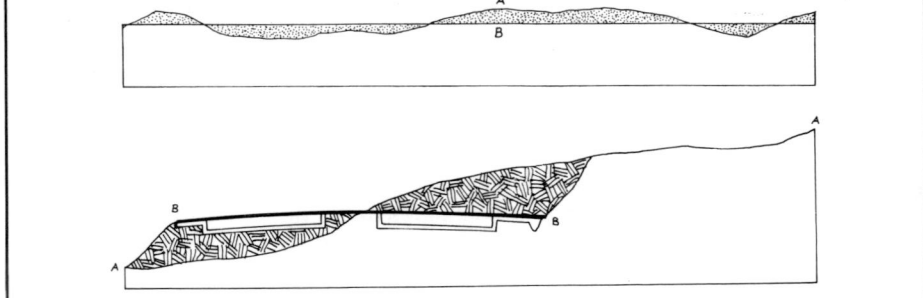

Box 2 General criteria for location of roads

Criteria	Explanation
• meet fixed point constraints	route has fixed starting, ending and particular crossing points, e.g. for rivers and railways
• remain within corridor of traffic demand	route must remain near centres of attraction (e.g. cities) otherwise traffic demand falls
• obey constraints of geometric design standards	horizontal/vertical curvature and width of road depend on predicted level of traffic and type of road being planned
• respect topography	if terrain is particularly hilly or mountainous, design needs to be adjusted to minimise rise and fall of road, and balance **cut** and **fill** (see Figure 2)
• take geology and ground conditions into account	some locations may be more expensive than others, e.g. if they require blasting through hard rock or crossing flood plains
• harmonise with the landscape	roads should add (or make no difference) to the quality of the view, not lessening its visual appeal
• take physical constraints into account	route should avoid power lines, settlements, historic buildings, military areas, farms, etc

so this difference has been highlighted by exaggerating the scale of the cross-section. Where the ground is lower than the road, soil has to be brought in to raise the level and this material is described as **fill**. Conversely, where the ground is higher than the road, soil has to be removed, and this is referred to as **cut**.

An economic design will ensure that cut and fill roughly balance over the whole length of the scheme. If there is insufficient soil it has to be excavated from elsewhere, thus creating **borrow pits**. And if there is too much soil it has to be dumped, creating **spoil tips**. Even when cut and fill broadly balance over the length of a road it may be necessary to create intermediate borrow pits and spoil heaps to reduce the costs of moving the soil. However, with good landscaping, these pits and heaps can be turned into attractive features, such as small lakes or hills.

Design standards Speed is the most significant factor affecting road design – both **design speed** and **operating speed**. The design speed is the speed below which vehicles travel safely, and depends on things like road curvature, gradient and sight distance. The operating speed, on the other hand, is the overall speed at which a vehicle travels in good weather and under the usual traffic conditions. Because traffic volumes limit the speed at which individual vehicles can travel, the operating speed is usually less than the design speed.

By convention the design speed is normally taken as the speed below which 85 per cent of drivers travel when the road is wet, and not busy. For a new road the design speed is estimated, taking into account factors such as bendiness, visibility, verge and road width, and other potential layout features. Obviously a motorway will have much higher design speed than a narrow single carriageway road with frequent junctions.

Visibility To overtake and stop safely a driver needs to be able to see the road ahead. The stopping sight distance (**SSD**) is the sum of the distance travelled during the thinking plus reaction time (usually taken as two seconds) before the brakes are applied, and the braking distance proper. You probably remember learning about thinking, braking and overall stopping distance in the Highway Code. And the SSD is simply the overall stopping distance, taking into account weather and road conditions.

Stopping sight distances increase significantly as the speed of the vehicle rises. The SSD at 140 kilometres per hour is about three times that at 70 kilometres per hour, and the SSD will also increase or decrease according to whether the road is going downhill or uphill. Curves and gradients therefore have to be planned to allow the driver to see his SSD.

Full overtaking sight distance (**FOSD**) is required when overtaking vehicles in the opposing lane on new or improved single carriageways. However, because of other constraints, it is not possible to maintain FOSD over the whole length of a road, and this is why carriageway markings warn motorists when it is dangerous to overtake. The FOSD is substantially greater than the SSD and as design speeds increase it becomes more difficult to provide.

BUILDING ROADS

> **Box 3 Full overtaking sight distance**
>
> The four main components of FOSD are:
>
> *The perception reaction distance*
> This is the distance travelled by the vehicle while the driver decides to overtake.
>
> *The overtaking distance*
> This has three parts: when the overtaking vehicle moves out into the opposing lane and alongside the slower vehicle, when the vehicle moves safely ahead of the slower vehicle, and when the overtaking vehicle moves back into the nearside lane, leaving a safe gap between it and the overtaken vehicle.
>
> *The closing distance*
> This is the distance travelled by the oncoming vehicle while the actual overtaking manoeuvre is taking place.
>
> *The safety distance*
> This is the distance required for clearance between the overtaking vehicle and the oncoming vehicle when the overtaking vehicle has returned to its own lane.

Structure of roads

A road has three main functions:

- to carry the traffic load and distribute its weight over the soil beneath
- to provide a surface that will be safe and strong enough to resist the effects of traffic and weather
- to prevent rainfall and surface water from entering the underlying soil and thus help to maintain soil stability.

There are three main types of road pavement: **flexible, rigid** and **composite**. In flexible construction the surfacing and road base materials are held together with bitumen. In rigid construction the road base and surfacing are in the form of a concrete slab in which cement is the binder. And in composite construction the upper pavement layers are made of bituminous material and are supported on a cemented road base.

To determine the thickness of the various elements of a road structure, the engineer has to ask the following questions:

- How heavy is the traffic loading?
- How strong are the materials making up the surfacing, road base and sub-base?
- What is the bearing strength of the subgrade soil?

INSIDE SCIENCE

> **Box 4   Major elements in road structure**
>
> **pavement**   consists of surfacing, road base and sub-base
> **surfacing**   provides a safe and comfortable surface for traffic
> **road base**   distributes weight of traffic over sub-base or formation
> **sub-base**   one or more layers of material supporting road base
> **formation**   surface of subgrade after completion of any earthworks
> **subgrade**   upper part of soil which supports loads carried by pavement

Road building materials

Figure 3: *(a) Rigid pavement (b) Flexible pavement*

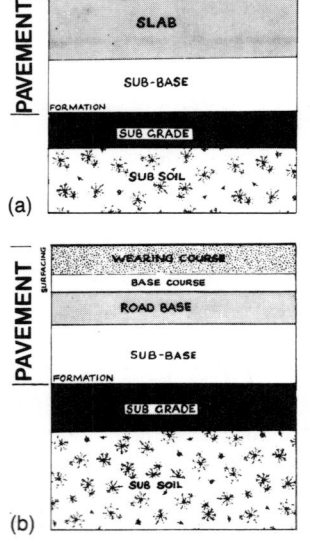

The pavement has to be thick enough to spread the load, and this thickness will depend on the material used. Technically this property is known as the material's **shear strength**. Low shear strength material is likely to be cheaper and more easily available but a greater thickness is required, so it is not always the most cost-effective option.

A flexible road pavement is generally a carefully graded mixture of gravel, sand and clay. The particular combination of strength and flexibility is determined by the mix of materials used.

In rigid road pavements, the concrete slab frequently combines base and surfacing, and so a sub-base is not often needed. A concrete slab can spread a load better than a similar thickness of flexible base but this does depend on the microscopic structure of the concrete.

However, careful preparation of the subsoil (and the sub-base, if there is one) beneath rigid slabs is more important than for flexible bases. And reinforcement of the slab is often required where the subsoil support is uneven.

Most road surfacings contain a bituminous binder. Bituminous materials include naturally occurring asphaltic bitumens, petroleum distillates (bitumens) and coal distillates (tar and pitch). The surfacing may be more than one layer thick, the top

130

Figure 4, *above: Rigid pavement construction;*
Figure 5, *right: Flexible pavement construction*

layer providing a good surface for traffic and waterproofing the underlying layers and subsoil. It should also prevent skidding under all weather conditions.

When designing the strength of roads, most attention is paid to the number of heavy vehicles. This is because of the relationship between axle load and the damage caused to road structures. This damage is proportional to approximately the fourth power of the axle load, so if the axle load doubles, the damage factor increases 16 times (2^4). If the axle load trebles, the damage factor increases by 81 times (or 3^4). As the axle load of a commercial vehicle may be five or ten times that of a car, it is at least 625 times more damaging, so the effects of ordinary cars on the road structure can be largely ignored.

Maintenance

On roads which carry a large amount of traffic, all road surface materials with a bituminous base gradually consolidate and become polished. The effect on concrete is not quite so marked. However, the net effect is to reduce the friction between the tyre and the road surface, and thus increase the risk of skidding, particularly when the surface is wet. Periodic maintenance is therefore important to keep the texture of the road surface reasonably rough.

The approach to maintenance depends on the road's 'whole life cost'. The object is to minimise the total cost (of both initial construction and maintenance) over the road's design life, and there are a number of possible strategies for doing so. Thus, in some circumstances, it may be cheaper to build the road to a lower specification and periodically reconstruct it over the next 20 to 40 years. Alternatively, it might be cheaper to build it to a very high specification. Then, apart from routine maintenance (cleansing of drainage channels etc), it would only require periodic upgrading of the running surface. Which of these alternatives is best depends on the type and location of the road (urban or rural), the overall volume and composition of traffic, the likely level of traffic delay or factors like the availability of alternative routes if the road has to be closed.

A modern highway probably requires periodic maintenance two or three times over its 40-year life. So, after the first ten years of operation, each year will see a section of road having some form of periodic maintenance, lasting two to three months, and requiring some restriction on traffic flow in the form of lane closures or contra-flows.

Unfortunately, it is often not possible to restrict maintenance work to non-daylight hours or carry it out at times of the year when there is a risk of frost or snow.

Safety fences

We place crash barriers (or safety fences) on the central reservation of a motorway to prevent out-of-control vehicles from crossing on to the opposite carriageway. These fences are usually fairly flexible, to absorb some of the energy of an impact. The hope is that when a vehicle hits the barrier its rear will rotate towards the fence, slide along it and the vehicle will then be directed back in the original direction of travel. The effectiveness of this will depend on the angle of approach and the vehicle's speed and weight. The greater these are, the more likely it is that the vehicle will cross the barrier or overturn. Present-day safety fences are designed to cope with the impact of a car weighing 1.5 tonnes striking at 113 kilometres per hour (70 miles per hour) at an angle of 20 degrees to the line of the fence. Such a fence will not, however, prevent heavy commercial vehicles from crossing on to the opposite carriageway, because they cannot absorb the total energy of the impact.

To be effective, safety fences have to be built exactly to specification with the safety beam mounted at a precise height above ground level. In the past the posts to which the beam was attached had so-called shear bolts at the same level as the beam. When a vehicle crashed into the fence these bolts would fracture, allowing the posts to be knocked down (and so absorbing a lot of energy), but without noticeably affecting the height of the beam. Recently these shear bolts have been put lower to help prevent commercial vehicles rolling over when they hit the fence.

Safety fences are effective: less than three or four motorway accidents per year are caused by vehicles crossing the central reserve. When these accidents are analysed we find that the vehicles cross for a variety of reasons. Sometimes the impact is not one for which the barrier is designed (high-speed or head-on, for example). Alternatively, the safety beam may no longer be at the correct height following reconstruction or the beam material might be fatigued. Finally, the initial construction may be at fault: poor foundations may result in the supporting posts coming out of the ground rather than fracturing at the shear bolt. This means that the beam is dragged down towards the ground and the vehicle passes over.

Road building in the future

Modern highway engineering is continuously changing and developing, with new technology in the form of Computer Aided Design (CAD), improved and faster methods of construction and new materials. Many developed countries have perhaps come to the end of their major road construction programmes and now face the task of maintaining and reconstructing older parts of their network. However, the young engineer can still look forward to the challenges of modern highway engineering in the many less developed countries of the world.

ARTIFICIAL INTELLIGENCE
The electronic brain?

Igor Aleksander and Helen Morton

This chapter investigates:
- the meaning of artificial intelligence
- what computers can do
- the use of logic in programming
- playing games with a computer
- solving problems with a computer
- the knowledge-based approach
- the 'human' computer
- making a computer 'see'
- machine intelligence versus human intelligence

What is artificial intelligence?

Many people would describe a computer as 'an artificial brain'. They know that computers can perform calculations very efficiently but they may not have thought about a computer's other abilities. In order to play games and plan tasks, a computer also performs sequences of logical operations very quickly. This area of computing is known as **artificial intelligence**, and it is the speed at which these logical operations are performed that makes it appear to be like real intelligence. In fact the individual steps involved are very simple but the programmer can combine them in different ways to make the computer perform very complex tasks.

What can computers do?

In this chapter we are going to look at what someone sitting at a computer keyboard can and can't expect to achieve, but in order to understand this we need to find out a bit more about how a computer works.

Programs are written in computer languages such as BASIC, PASCAL or PROLOG, and different languages are best for different things. But as we are only writing 'thought programs' we shall use simple instructions in English, rather than a particular com-

> **Box 1 What is a computer?**
>
> A computer consists of three main parts:
>
> - a **processing unit** that can perform tasks involving arithmetic or logic (such as working out whether one number is the same, greater or smaller than another)
>
> - **memory** which stores **data** and a **program** or list of instructions that are executed one by one (an instruction contains information that tells the processor where to find data and what to do with it)
>
> - **input** and **output devices** which can receive instructions from the user or programmer (often a keyboard or a magnetic disk) and output the results (usually a printer or a TV screen)

puter language. For example, here is an instruction:

```
Print the answer to 12 x 6
```

If this instruction were translated into a computer language, the computer would show (or print) 72 on some part of its screen (or printer). The first appearance of intelligence can be added by means of a slightly more elaborate series of instructions.

```
Print 'Please type in a number'
Read this number and call it A
Print 'Please type in another number'
Read this number and call it B
Print 'The product of your two numbers is'
Print the answer to A x B
```

If we were to 'run' (or use) this program then we would enter numbers when the computer requested them, for example:

```
Please type in a number 12345
Please type in another number 6789
The product of your two numbers is 83810205
```

135

Not only is the machine able to do arithmetic at very high speed, but it also appears to communicate in a 'human way' with the user. It does this by repeating precisely the form of words it was given by the programmer. This small program demonstrates two things that we can expect from a computer: the first is its obvious ability to do arithmetic and the second is its capacity to 'understand' what one might call housekeeping commands (or instructions) such as 'print' and 'read'. This latter ability is limited, and those writing instructions have to know which housekeeping commands are available in a particular language. For example, you might try to use the following instruction:

Whistle the tune of 'God Save The Queen'

However, the computer would probably print 'illegal command' because 'whistle' was not one of its housekeeping commands.

Experienced programmers do not worry about this kind of limitation. They would connect a sound synthesiser to their computer, and use the word 'whistle' to make the computer look at a separate mini-program called a **subroutine** or a **procedure**. This subroutine would contain sequences of instructions that the computer does 'understand' and which drive the synthesiser in order to produce the desired whistle.

Algorithms It was the British mathematician and philosopher Alan Turing who, in 1936, showed that the only limits to what can be computed lie in the programmer's ability to express the procedure as a simple series of clear steps. These steps are known as an **algorithm**, and Turing showed that the simplest possible machine (known as a Turing machine) which just copies, prints and reads symbols on a long tape can execute the most complex series of steps, as long as they are clearly stated. Turing's discovery was a superb exercise of the imagination as the discussion was based entirely on fictional machines. The first computer, as we now know it, was only designed ten years later by John Von Neumann.

So, from here on we shall often express what a computer does simply in terms of an algorithm, knowing that it can be programmed in an appropriate computing language. For instance, the algorithm for our previous example might read:

```
'Ask for two numbers and print the product'
```

This of course assumes that the computer has other algorithms that define 'ask', 'print' and 'product'.

The use of logic in programming

The phrase 'Artificial Intelligence' was coined by John McCarthy in the late 1950s, and it marked the realisation that a computer could perform operations based on logic as well as arithmetic. For example, the following is a typical program that depends on simple logic:

```
Print 'Do you have a runny nose? (type yes or no)'
Store ANSWER 1
Print 'Do you have a temperature? (type yes or no)'
STORE ANSWER 2
If ANSWER 1 = yes and ANSWER 2 = yes then print
    'You may have influenza'
If ANSWER 1 = yes and ANSWER 2 = no then print
    'You may have a cold'
```

Whether or not the medical diagnosis is correct, the above program shows the use of **operators** such as 'if', 'then', and 'and' (as well as 'not' and 'or', not used above), which extends the power of computers beyond arithmetic and into the area of logic. Indeed **expert systems**, which are intended to capture areas of human knowledge, are largely based on logic in a similar way to the example given above.

Playing games with a computer

In 1950 Claude Shannon, of the Bell Telephone Laboratories at Murray Hill in the USA, suggested that a computer could play chess by using logic to search through the possible moves and choosing the most promising one. To describe this process, we shall look at a somewhat simpler game, remembering that the same principles could be applied to the more complex situation of chess.

INSIDE SCIENCE

This game is a version of 'join the dots' in which dots are arranged in a square grid. Each of two players can join two neighbouring dots to make one side of a square. Every time the players manage to complete a square they claim it, and choose whether or not to have one more move. Then the other player takes over. At the end of the game, when all the squares are closed, the player with the most squares wins.

Our version of this game is simplified to the extreme and has only six dots arranged as shown:

. .

. .

. .

A pair of game sequences is shown in *Figure 1*. In the first, the starting player loses as his opponent has captured both squares (indicated by the grey marking). In the second game, the starting player manages to win (his squares are shown as black).

For a computer to play this game it is necessary to take alternative moves from the keyboard, and for the computer to produce its own moves based on some logical playing strategy. One of Shannon's algorithms involves using the rules of the game stated as logical 'If-Then' statements of the kind we saw above, to generate all the moves that can be taken, looking several steps ahead. For our simple game there can only be seven steps in the game, so

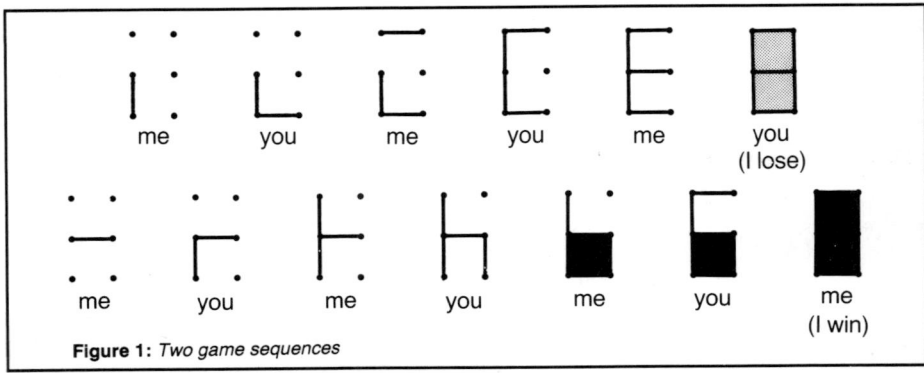

Figure 1: *Two game sequences*

it is possible to list all the significantly different moves at each stage. For example, the following moves are not significantly different because they are mirror images of each other:

whereas these two are very different:

Figure 2 (on the next page) shows all possible moves. The essence of the algorithm is that the computer does not work out the best move at each stage. Instead, it finds the best outcome at the furthest moves it can simulate, and then makes the move most likely to achieve that outcome.

A further advantage of this algorithm is that it is not necessary to play the game to the end in order to calculate the necessary values of the playing moves. One simply drives it as far as its speed and the size of the computer's memory will allow. All that Shannon's algorithm requires is that any game state can be evaluated in terms of its likelihood of leading to victory for the machine. The machine then evaluates those states at the furthest level, and backs up from there. The backed-up values are used in preference to the first level values as they contain more information about the future prospects of the move.

In games such as chess it is necessary to write a set of rules which allow the evaluation of the states to be carried out. For example, a state which threatens the opponent's king (mate) would be highly valued, while one that sacrifices one's own piece would be given a low value. Such rules are known as **heuristics** (from the Greek word *heurisko*: to find; that is, rules that are known to help in finding the answers but which may not have an immediate mathematical explanation). Shannon also proposed efficient ways of cutting out branches of the search tree on the basis that, having achieved a particular value in a partial search, the search need not continue.

Game-playing has proved to be highly marketable: a great variety of chess, backgammon, bridge, othello and similar games programs are widely available. There are also prestigious computer chess tournaments where programmers pit their wits against one another in a battle of heuristics. But at the end of it all, a chess master is still only rarely beaten by a program. People play chess in a very different way from computers. They recognise winning trends, opponents' styles and so on. They are too intelligent to rely on an exhaustive search of the massively increasing number of moves that need to be evaluated as one looks ahead in the game.

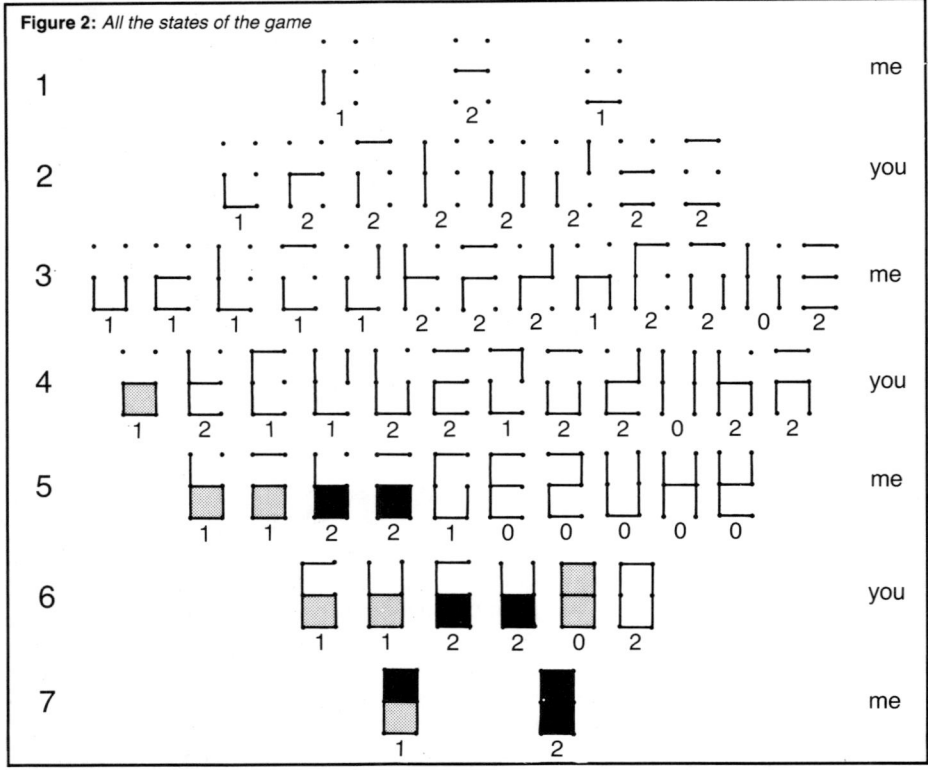

Figure 2: *All the states of the game*

ARTIFICIAL INTELLIGENCE

Box 2 The 'MINI-MAX' algorithm

Shannon's algorithm, called the MINI-MAX (for reasons that will become obvious) starts with some form of evaluation of the *furthest* level of moves. In our example, at level 7 we have a winning move for the machine ('me') in which both squares have been captured. We give this move a value corresponding to the number of squares that have been captured: 2. Similarly, the other move at level 7 where only one square belongs to the machine is given the value 1. At level 6 there is a winning move for the opponent, and this is given the value 0, because it is least valuable to the machine. Having been given the logic for evaluating these furthest moves by the programmer, the program must calculate values for all the other moves using a process of **backing up**, assuming that the opponent will take the move which minimises the machine's advantage.

In the closing stages of this game, there is virtually no choice in going from one move to another, and this lack of choice simplifies the backing-up process. We can therefore look at a small part of the labelling process at levels 2, 3, and 4 which illustrates the 'MINI-MAX' algorithm at work (*see Figure 3*). This form of analysis tells us quite a lot about the game. For example, assuming that both players play the best moves, the starting player cannot lose.

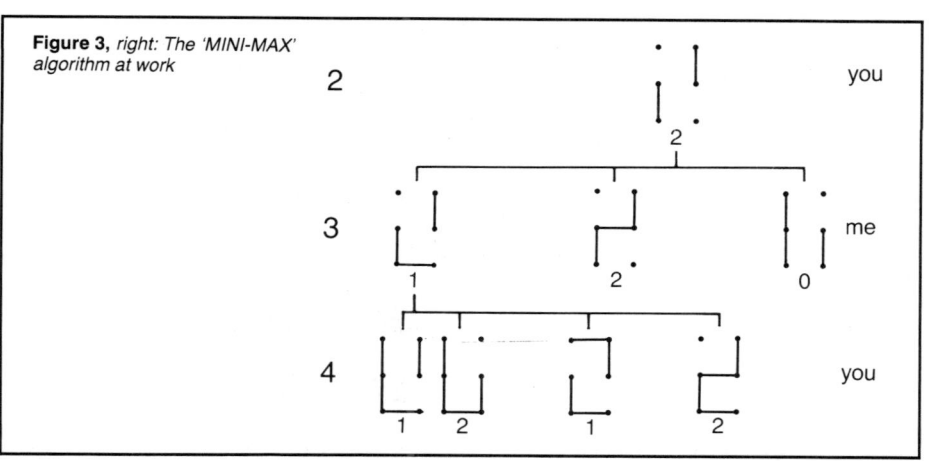

Figure 3, *right: The 'MINI-MAX' algorithm at work*

Solving problems

In the 1960s any activity at which humans were good and machines bad became fair game for supporters of artificial intelligence. At this time computers often solved puzzles by simply working through all possible choices. The next development was **means-ends analysis**.

This type of 'means-ends' program would assume that all problems could be formulated by **goals** (such as 'I am in London and need to get to New Jersey'), **sub-goals** (such as 'Kennedy airport in New York') and means (such as air travel, taxis, buses, etc) plus information about which means are suitable for which goals. The machine would then search for solutions against some objective such as minimum cost or least time.

Now, however, we want to develop more general systems. Expert or **knowledge-based** systems have raised problem-solving to the more general level of **automated reasoning**, in which the computer has a database of facts and rules about the relationships between these facts. To illustrate the development of this technology we shall look at an example concerned with the stacking of three blocks: a large one, a medium one and a small one.

Figure 4 shows all the possible ways of stacking these blocks. The lines, or links, show how these arrangements could follow one another by moving just one block. This is similar to the game-playing program for 'join the dots' in that each stacking arrangement can be seen as the equivalent of a state

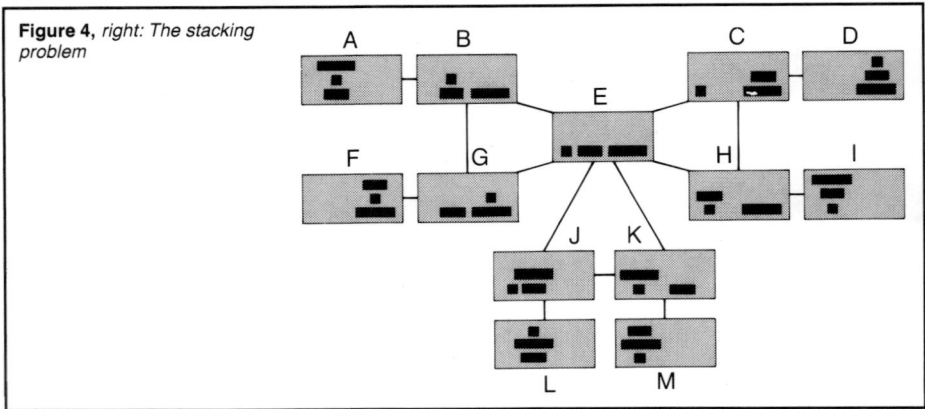

Figure 4, *right: The stacking problem*

of the game. Each stacking variation is therefore a state of the stacking task. However, unlike 'join the dots', the links between states go both ways, as any action can be reversed. So, rather than having a starting state and a winning state, we express the problem in terms of a starting state and a target state in which the blocks are stacked in a particular way. The aim of the program is to find a path between the two.

Box 3 Stacking the blocks

Suppose we want to go from state F to state D. One way to solve this on a computer is to store the states shown in *Figure 4* and the links between them. The starting state leads to a number of 'children' states which are examined to see if one of them is the target state. If not, the children become the parents and new children states are investigated. A simple rule is added to avoid visiting the same state twice: the new 'children' cannot include states that have already been 'parents'.

So, if F is the starting state, the first child is G, the next children are B and E, and the next A and C – which is followed by D, our target state. We can now trace the sequence that leads to the solution: F-G-E-C-D. But we have not translated this sequence of moves into instructions that a computer could use to move from state F to state D. To provide these instructions, we need to label the transitions between states. For example, the transitions between state F and G can be written in shorthand as follows:

F G: M, on (S), on (table).

This means: if the change is from F to G then move the medium block (M) from being on the small block (S) to being on the table. So the solution F-G-E-C-D becomes (in 'algorithmese'): move M from on S to on table; move S from on L (large) to on table; move M from on table to on L; move S from on table to on M.

Solving a problem by going from the start to a target state is known as **forward deduction**. It is equally possible to work backwards from the target, or, even better, both forward from the start and backwards from the target state. In this case the search is complete when the two searches have at least one state in common. However we have recently come to regard this way of searching 'state trees' as often too particular a way of solving problems.

The knowledge-based approach

In the last example all the states of the problem were stored, with the links (or transitions) labelled, in a **database**. The database contained **records** describing each state and its links to other states. So this knowledge-based approach clearly involves storing the rules that define a problem as well as some facts about the individual states. A system such as this avoids storing a complex tree structure and provides a more general way of retaining knowledge. And this is the type of approach that is used in expert systems.

A program that finds the appropriate rules, applies them and keeps track of the current states in the search is called an **inference engine**. This is because its task is to 'infer' the next state, having been given a description of the current state.

It is precisely this combination of rules, facts and an inference engine that is at work in expert systems. And these systems can be applied to any task for which the facts and rules can be clearly stated (for example, oil prospecting, medical diagnosis, legal advice, managerial advice, design and so on). There are also **expert system shells** which provide an inference engine and a convenient way of stating rules and facts so that users can build their own knowledge base. A specialised programming language, PROLOG, allows users to express everything in terms of this type of logic statement, by converting statements into appropriate logical rules. PROLOG therefore combines an inference engine with logical functions as well as the traditional mathematical functions of a computer language.

The 'human' computer

Present-day computers are designed to carry out simple instructions at very high speed but they still find it difficult to do some things that humans find easy. We once gave a lecture to young people on computer vision, and a six-year-old girl said afterwards, 'Computers must be very stupid if they can't see. I don't have to think to see, but I do have to think to do my sums.' Indeed, good vision, speech and language systems are still very much in the future. Quite a lot of progress has been made, using logic techniques to check the grammatical validity of sentences: after all grammar is just a collection of logical rules. But capturing the meaning (the semantics) of a sentence has proved more difficult. What success has been achieved, has taken place in very restricted contexts, like our stacking scheme. It is conceivable that a human instruction at state B, such as 'put the small block on the large block' could be analysed by identifying the parts of speech from grammatical rules, and then looking up the actions implied. But developing this into more general applications is not a trivial task.

In vision systems, limited success is possible with logical rules, as shown in *Figure 5*. This example suggests a straightforward way of identifying simple images, using logic to identify features and surfaces.

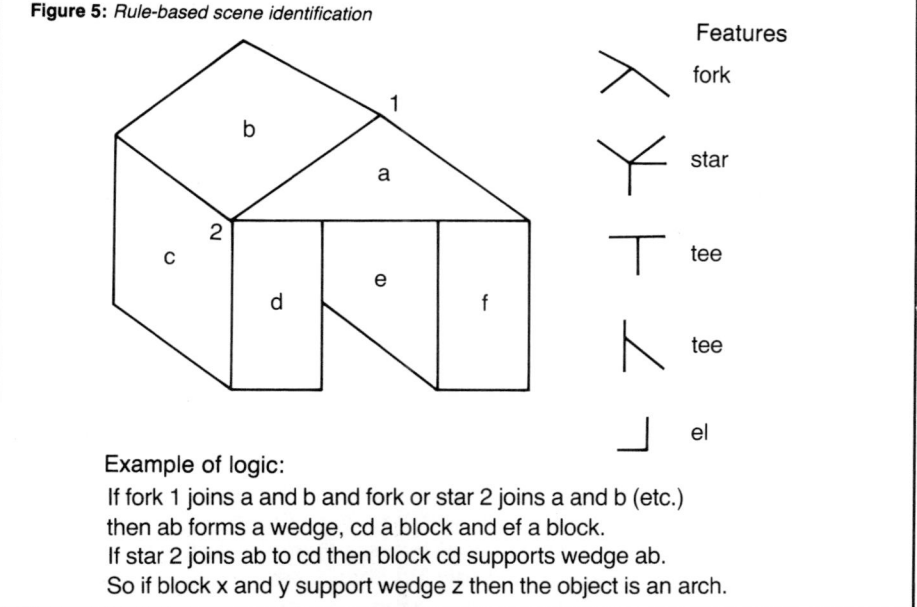

Figure 5: *Rule-based scene identification*

Example of logic:
If fork 1 joins a and b and fork or star 2 joins a and b (etc.)
then ab forms a wedge, cd a block and ef a block.
If star 2 joins ab to cd then block cd supports wedge ab.
So if block x and y support wedge z then the object is an arch.

Unfortunately, however, this system is difficult to apply to scenes which have badly defined features (like the fuzzy images which may be produced by a TV camera in a factory). To overcome this problem, many industrial vision processes have been developed on the basis of mathematical formulae, rather than logic. These 'mathematical' vision systems work well in factories when they are used for specific tasks such as quality control and helping robots to 'see'. However, they are only applicable to particular problems (e.g. detecting cracks in images of bearings), and are ad-hoc solutions to ad-hoc problems, rather than applications of artificial intelligence.

So, whither the artificial human?

Making a computer 'see'

Computer pioneers saw the present-day digital computer as just one example of what could be done with electronics. This type of computer now dominates the industry but an alternative is to develop networks of electronic models of **neurons** (the cells in the brain). Such a computer is described as a **neural net**.

The essential difference between neural nets and conventional computers is that neural nets build up knowledge through learning and being trained, while computers need to be programmed on the basis of an algorithm. The neural alternative was almost forgotten a few decades ago, but now things have changed. Largely because of the limitations of having to dream up the rules necessary to drive algorithms and the greater power of modern computer chips and new optical technology, neural computing is becoming a practical possibility.

There are three processes which take place in any vision machine, including the human brain: a goal has to be set, an algorithm has to be defined that links input to output, and that algorithm has to be implemented. There are also four distinct parts of a framework (shown in *Figure 6*) which link the entry of a light signal into a vision box to its eventual description as a meaningful object.

As *Figure 6* shows, when working with computers we base our methods on what we know to be the processing steps taken by the neural

Figure 6: *David Marr's approach*

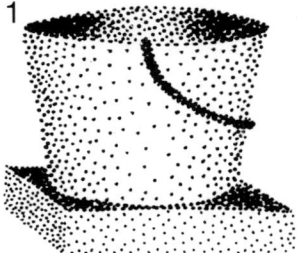

Raw image: local intensities

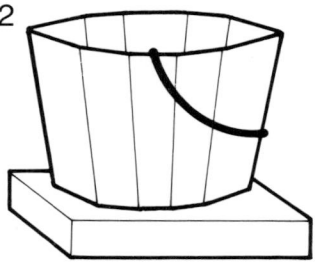

Primal sketch: Changes between intensities, e.g. edges and blobs

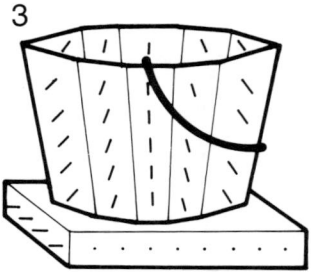

2½-dimensional sketch
Primal sketch + knowledge of lighting, surface reflection, etc.
= local orientations and relationships

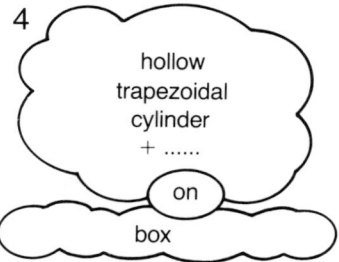

3-dimensional model
knowledge of the world:
hollow, trapezoidal cylinder
+ part ellipse on box
= bucket on slab

pathways of the brain. The first part of this framework is the pattern of intensities which is detected in the human brain by the retinal cells which respond to both light and colour intensity. Deeper in the visual system there are cells which respond to an edge or a blob, which leads to the second part of the framework: the primal sketch. Here, changes in the intensity patterns are used to distinguish lines and other important features of the two-dimensional image.

At the third stage, it is possible to add 'half a dimension' to the computation so far. Knowledge of lighting and surfaces is used automatically by the brain, and this knowledge makes it sensitive to local positioning and relationships in order to decide how things fit together. This is called the $2^{1}/_{2}$-dimensional sketch. The information provided by the $2^{1}/_{2}$-dimensional sketch forces the computer to decide what the image is, and this meaning is selected from the machine's known shapes. For example 'hollow trapezoidal cylinder' is one meaning that can be selected. Further properties such as 'mouth up' and 'ellipsoidal handle' can lead to the conclusion that a bucket is present!

To use these recognition techniques successfully, we have to teach the computer about different shapes and objects. In other words the computer needs to be trained.

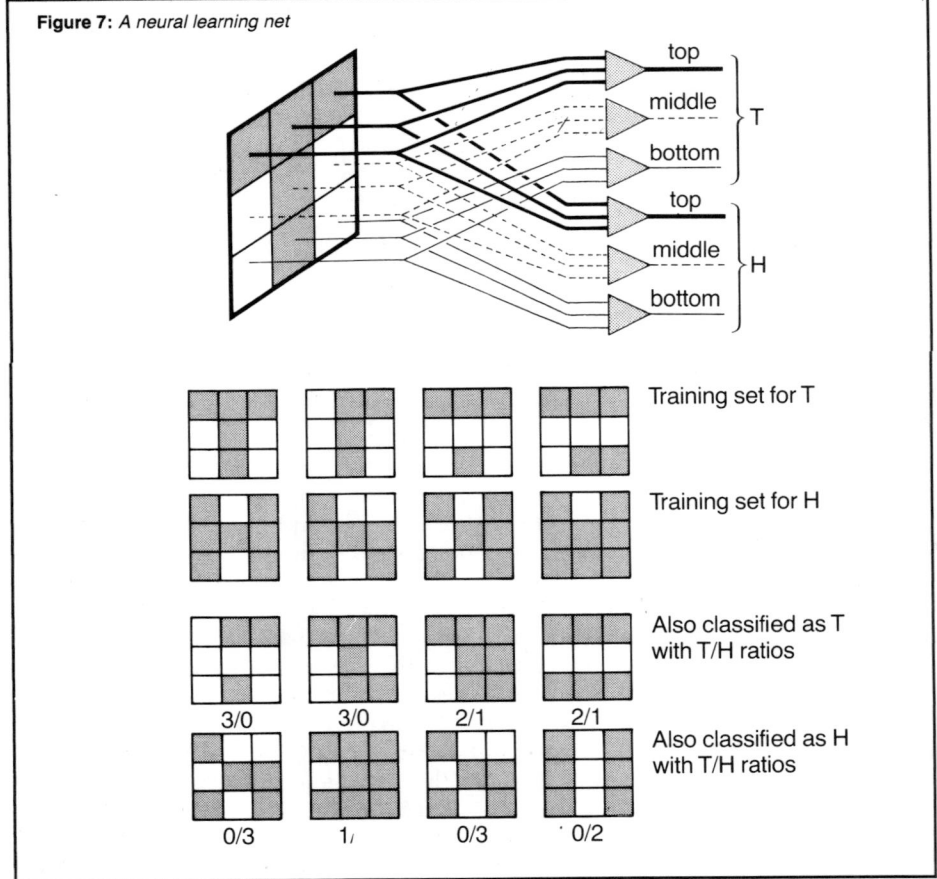

Figure 7: *A neural learning net*

Figure 7 shows an example of the way in which a very primitive net can generalise from training examples. A neuron is a device which learns to recognise patterns at its inputs, and the triangles represent neurons which can be thought of as tiny memories. Each neuron in this simplified example receives three inputs from the image. During training the neuron learns to take these inputs and produce a 0 or a 1 at its output. In a neuron within the brain the data is stored by variations of chemical activity at its input, known as the **synapse**.

The model presented in *Figure 7* comes from the author's own research laboratories and shows a highly stylised but practical way of making neural nets. Other models, which resemble the neuron more closely, tend to be harder to simulate.

Returning to *Figure 7*, we see a three by three grid with an upper and lower group, each consisting of three neurons. We wish to train these neurons to recognise the letters 'T' and 'H' respectively. Initially, whatever the image, the neurons give a 0 on their outputs. By using the training patterns, the top three learn to respond with a 1 to the letter 'T', and the lower three learn to respond with a 1 to the letter 'H'.

Having learnt how to recognise 'T' and 'H' the neurons will then attempt to 'see' these objects in other images, responding more strongly (with more of its units saying '1') on the T-side to patterns it 'thinks' look like Ts (and similarly for Hs). A vision system called the WISARD which uses up to half a million, eight-input logical neurons has been developed on these principles.

Machine intelligence versus human intelligence

By now, it should be clear that machine intelligence is not the same as the human variety. The living animal can teach us many techniques and frequently remains better at tasks that may be said to require intelligence. But the engineering value of what has been achieved in artificial intelligence should not be underestimated. Its main benefit lies not in making machines that replace humans, but machines which, because they can behave 'intelligently' in their own terms, allow their users to make better use of them.

List of contributors

Chapter 1 *Nuclear energy*
Dr Helen ApSimon graduated from Oxford University and followed a doctorate in astrophysics with consultancy work in the nuclear energy field. She now heads an Air Pollution Research Group which has been involved in extensive analysis of the Chernobyl accident.

Professor Tony Goddard worked for the Atomic Energy Authority for several years before joining Imperial College, London. He is now Professor of Environmental Safety at Imperial College and his particular research interests are radiation protection, fusion technology and the environmental impact of nuclear power.

Chapter 2 *The energy 'crisis'*
Dr Richard Marshall gained his doctorate at Sussex University and now lectures at Keele University, where he has developed a very successful undergraduate Energy Studies course. His publications include studies of the efficiency of heating systems and the potential of solar power systems.

Chapter 3 *Test tube babies*
Dr Virginia Bolton read zoology at Oxford University, and gained her doctorate at Cambridge University. She remained at Cambridge to do post-doctoral research in IVF, and is now Senior Embryologist and Honorary Lecturer at the Assisted Conception Unit, in the Academic Department of Obstetrics and Gynaecology, King's College School of Medicine and Dentistry, London.

Dr John Parsons graduated from medical school in Dundee. After working as a General Medical Officer in Central Africa, he trained as an obstetrician and gynaecologist in the United Kingdom. He is now Senior Lecturer in charge of the Assisted Conception Unit, in the Academic Department of Obstetrics and Gynaecology, King's College School of Medicine and Dentistry, London.

Professor Stuart Campbell studied medicine at Glasgow University and trained in obstetrics and gynaecology in Glasgow and London. He has an international reputation in the field of perinatal research, and in 1976 he was appointed as Professor of Obstetrics and Gynaecology at King's College School of Medicine and Dentistry, London.

Chapter 4 *Pollution*
Dr Nigel Graham graduated from Cambridge University and completed his doctorate at Imperial College, London. He has worked for Thames Water Authority and two consulting engineering companies, and now lectures at Imperial College. His main research interests are the supply and treatment of water.

Chapter 5 *Lasers*
Dr Robert Eason studied applied physics at London University. After two years at Bristol University he did research on lasers at York University, where he received his doctorate in 1982. He has lectured at Essex University and is now at Southampton University.

Chapter 6 *Plastic*
Dr Ray Smith worked for a major international pharmaceutical company for ten years and graduated from the Royal Society of Chemistry in 1979. He received his doctorate from Manchester University, where he did research on water-soluble polymers, before starting his post-doctoral research on the degradation of polymers at Liverpool University. He now lectures at Queen Mary College, London.

Chapter 7 *Healthy eating*
Dr Alan Stewart trained at Guy's Hospital, London. He is a member of the Royal College of Physicians and Medical Adviser to the Women's Nutritional Advisory Service. His major research interests are nutrition and the role of nutritional factors in pre-menstrual tension.

Chapter 8 *The changing Earth*
Dr Hazel Rymer studied physics at Reading University, then did geophysical research for her Open University doctorate. She is a Chartered Physicist and Geological Society Fellow. Her research interests are the use of gravity and magnetic methods to monitor and model volcanic activity and hazard.

Chapter 9 *Building roads*
Dennis Gilbert gained a BSc in civil engineering from London University and an MSc in transport from Imperial College, London. He is a Chartered Engineer and a member of several bodies with interests in highways and transportation. He spent a number of years with local authorities designing highway schemes before joining Imperial College as a lecturer. His main research interest is road traffic and the environment and he has published widely on this and other subjects.

Chapter 10 *Artificial intelligence*
Professor Igor Aleksander trained as an electrical engineer in South Africa and completed his doctorate in the United Kingdom. He has lectured at London, Kent and Brunel Universities, and now holds the Kobler chair in the Department of Computing at Imperial College, London. His numerous publications have covered fields ranging from the design of computing machinery to the modelling of intelligent behaviour in living organisms.

Dr Helen Morton lectures in psychology at Brunel University. Her main research interest is artificial intelligence, particularly the design of knowledge-based systems that aid the process of counselling.

Index and glossary

Many of the terms which appear in this glossary have more than one meaning. The definitions given here explain the way they are used in this book. Words in bold type within definitions refer you to another glossary entry.

ABS copolymer, 95
acid lakes, 52–5
acid rain, rainfall containing sulphuric acid and nitric acid, 54–5, 56
acidification, drop in pH of lake water, 52
acidity, hydrogen ion concentration of a solution; measured using the **pH scale**; pH value between 1 and 7, 53
additives (plastics), 88, 89–90, 93
Advanced Gas Cooled Reactor (AGR), 14, 19
alcohol consumption, 102, 104, 105–6, 107
algorithm, number of steps to be followed in sequence in order to solve a particular problem, 136–7, 138, 139
alkalinity, pH value between 7 and 14, 53
alpha-radiation, 17
aluminium pollution, 54, 55
amino acids, component **molecules** of **proteins**; compounds of carbon, hydrogen, oxygen, nitrogen and sometimes sulphur, 98, 99
amplification (optical), process in which one input **photon** produces more than one identical output photon, 70, 72–3, 74
analogue recording, recording based on a continuous range of signals, 78
andesite volcanoes, 115, 116
anti-nutrients, dietary compounds (such as coffee or alcohol) which prevent efficient absorption of nutrients, 101
anti-oxidants (plastics), 88

artificial intelligence, ability of a computer **program** to deduce information from other **data** and logical rules, 134
Atlantic Ocean, 116, 119
atom, smallest unit into which any substance can be divided; smallest particle of an **element** that can enter into a chemical combination; in each atom a **nucleus** is surrounded by **electrons**, and the number of electrons is equal to the number of **protons** in the nucleus, 8, 9–10, 83, 84
atomic mass, the average mass of **atoms** in a pure **element**; in an **isotope** it is close to the number of **neutrons** and **protons** in the **nucleus**, 11
atomic number, number of positively charged **protons** in the **nucleus** of an **atom**; normally the number of **electrons** around the nucleus also equals the atomic number, 10
automated reasoning, problem-solving in an **expert system**, 142

backing up, part of a game-playing **program** where values of board states are transferred from a move ahead in the game to an immediate move, 141
bad quality river, river with badly polluted water which has an offensive smell and appearance, 57
bakelite, 91–2
basaltic volcanoes, 116, 117
beaches, pollution of, 59–63
becquerel (Bq), unit used to describe the activity of **radioactive** material;

$1 Bq = 1$ disintegration (radioactive charge of an **atom**) per second, 18
beta-radiation, 17
binary, system of arithmetic or information coding using on/off signals only, 78
biochemical oxygen demand, oxygen required by organisms feeding on **biodegradable materials**, 58, 59
biodegradable materials, materials which can be broken down to simple chemical compounds by natural organisms, 57, 94–5
biodegradation, breaking down of materials to simple chemical compounds by natural organisms, 65, 94
biological nitrification, bacterial breakdown of nitrates, 65
birthmark removal, 75
blow moulding, blowing of molten plastic into a mould, 92
borrow pits, sites from which soil is taken for use as **fill** for road-building, 127
bowel cancer, 106
breast cancer, 106
buffering ion, dissolved substance which prevents pH falling by neutralising **acidity**, 55
bulk minerals, minerals required in large amounts to maintain structure and metabolism of the body, 100

cadmium pollution, 54
cancer, diet and, 106–7
carbohydrates, organic compounds which provide much of the energy in our diet; sugar is a carbohydrate, 98
carbon compounds, 86, 87

152

INDEX AND GLOSSARY

carbon dioxide laser, 74
carbon fibre composites, 93
carbon rings, structures formed by carbon **atoms** joined together to give pentagons (five carbons), hexagons (six carbons) etc, 87
cardiovascular system, heart and the blood vessels; the circulatory system, 105
Caribbean islands, 115
cervix, entrance to womb, 40, 47
chain reaction, reaction in which a **neutron** causes **fission** of a **nucleus**; from this fission neutrons are released which may cause more fissions so that the reaction is self-sustaining, 11, 12, 14–15
charge-discharge machine, 13
chemical energy, energy stored as the **kinetic** and potential energy of **electrons** in the chemical structure of substances; commonly released by burning, 25, 34, 35
chemical equations, 85
chemical reaction, interaction between **electrons** of different **molecules** or **atoms**, leading to filling and stabilisation of **electron shells**, 83, 84, 85
Chernobyl, 8–9, 14, 15–16
cholesterol, fatty substance (fat-soluble alcohol) found in all animal fats and in many animal tissues, 98, 105
chromosomes, fixed number of thread-like bodies, containing genetic material; present in all cell **nuclei**, 38–9, 41, 46
clingfilm, 90
coherence, degree to which light waves are 'in step' with one another, 67, 68
collimated light, highly directional beam of nearly parallel light, 67
compact disc (CD) players, 78

complex carbohydrates, insoluble non-sweet-tasting **carbohydrates** composed of large **molecules**; some form dietary **fibre** and cannot be digested, 98
composite materials, polymers with fillers added to change their mechanical properties, 92–3
composite pavement, upper **pavement** layers made of bituminous material supported on cement **road base**, 129
compression moulding, plastic pressed between the two halves of a mould, 92
Computer Aided Design (CAD), 133
computer games, 137–40
constructive plate boundary, boundary between two plates at which new oceanic **crust** is formed from **upper mantle**, 116
continental collision, collision between the continental parts of plates which cannot slide over each other and therefore buckle under the enormous pressure, 116
continental drift, process by which the formation of new oceanic **crust** slowly pushes continents apart, 116
control rod, rod which contains a strong **neutron** absorber, such as boron or cadmium; it is held partly inside the **core** of a nuclear reactor and its insertion or withdrawal enables the rate of reaction to be controlled, 12, 14, 15
coolant, liquid or gas pumped through the **core** of a nuclear reactor that takes the heat from **fission** and transfers it to a boiler where steam may be raised to drive the **turbine**, 12, 13, 14
coolant circuit, 14
copolymer, polymer made from two different **monomers**, 95

core, part of a nuclear reactor which contains the fuel, 8
coronary artery disease, build-up of **cholesterol** on inner walls of the two coronary arteries, restricting supply of blood to heart muscle, 105–6
covalent bond, bond formed by sharing **electrons** between **atoms** of a **molecule**, 84, 85
critical, state of a rector core when a steady **chain reaction** is maintained, 14–15
critically acidic lake, lake with water of pH below 4.5; plant and animal life seriously reduced, 53
crust, outer layer of the Earth's surface, 113, 116
cumulus cells, small 'nursing' cells surrounding eggs in the ovary and after **ovulation**, 41, 46
cut (in road-building), soil removed where ground level is higher than road being built, 127
cycle, one complete wave form, 69

data, information stored in a computer's **memory**, 135
database, stored **data** (e.g. facts, rules governing relationships and rules defining problems) used in **expert systems**, 144
dead or dying river, river with water so polluted that most natural forms of life have died out, 56, 57, 58
decay (radioactive), reduction in activity of **radioactive** material due to a proportion of the **atoms** present having changed ('disintegrated') into another form, 16, 18
delayed neutron, 15
design speed, speed below which vehicles travel safely on a road, 127, 128
destructive plate boundary, boundary between two plates at which crustal

material sinks downwards into the Earth's mantle, 115
digital recording, system of recording based on a system of on/off signals, 78
dissolved oxygen, oxygen dissolved in natural waters; essential for most animal and plant life, 57, 58
domestic heating efficiency, 33–4
donated eggs, 48, 49, 50
donated pre-embryos, 49, 50
double bond, equivalent of two **covalent bonds** between two **atoms**, 86, 87
Dounreay, 20
Drigg, 20
drilling, laser, 74–5
dual-ovenable trays, 90
dye laser, 75
dynamic hologram, hologram constructed using materials other than holographic plates and producing a moving image, 77
dynamo, device for converting **mechanical energy** into **electrical energy** by means of an electrical conductor moving across a magnetic field (or vice versa); sometimes called a generator, 26

Earth, history of, 111
earthquake prediction, 120
economic life, period after which it is no longer cost-effective to carry out repairs on a road, 124, 126
eggs,
 collection, 44–5
 donation, 48–9, 50
 frozen, 50
 normal production and transport, 38, 40, 41
elastic energy, form of **mechanical energy** stored in something that is compressed, e.g. a spring or a gas, 25, 28
electrical energy, 26, 27, 28, 29, 32, 35
electromagnetic energy, energy associated with electric and magnetic forces, 25
electromagnetic spectrum, range of electromagnetic **radiation** including visible light, 67, 68
electron, very light particle (compared with **neutrons** and **protons**) with a negative electric charge, 10, 17, 53, 83, 84, 85, 86
electron shell, orbit of **electrons** about a **nucleus**, 83, 84, 85
element, substance containing only **atoms** with the same **atomic number**, and so the same chemical behaviour, 83, 85
embryo, stage after implantation of fertilised egg in which potential fetal cells can be distinguished from those that will form **placental** tissue, 42
embryo freezing, 47
embryonic tissue, therapeutic uses, 51
endangered lake, lake with water of pH between 5.5 and 4.5 and consequent damage to plant and animal life, 53
endometriosis, development of uterine lining tissue in areas outside uterus, 43
energy, a property that measures how objects etc can interact and change each other, 24–5
energy conservation, 29, 30–1
energy crisis, 33, 34–5
energy efficiency, 33, 34
energy quality, property of any form of **energy** that measures how much of it could be converted into **mechanical energy** or 'work', 29, 31, 32, 33
energy storage, 35
energy transformations, 25–8, 32
engines, 31, 32
enrichment (nuclear), process whereby the proportion of fissile **atoms** (for example ^{235}U) in uranium is increased by artificial means, 19
entropy, a measure of **energy quality**, 29
enzymes, protein substances that determine which **chemical reactions** occur inside which cells, and alter the rate of biochemical reactions, 41, 100
epididymis, region next to testicle in which stored sperm become **motile**, 39
epoxy, thermosetting polymer often used in **composites** and adhesives, 93
essential fatty acids, linoleic and linolenic acids which must be supplied in the diet since the body cannot make them from other dietary components, 97
exercise, 105
expert system shell, computer **program** capable of **automated reasoning** which allows users to build their own **knowledge-based systems**, 144
expert system, data covering an entire area of detailed human knowledge which can then be manipulated by an **inference engine**, using a combination of rules and data, 137, 142, 144
extrusion, forcing of molten plastic through a nozzle to form pipes, tubes, sheets, etc, 91, 92

faecal coliform bacteria, 60
fallopian tubes, tubes which transport eggs from ovaries to uterus; normal site of fertilisation, 40, 41, 42
fast-breeder reactor, reactor in which the otherwise useless ^{238}U absorbs **neutrons** and is changed into fissile **plutonium**; this means that about 50 times more **energy** may be extracted, 20, 22
fats, 97–8, 106, 107, 108
feedback, return of some output from an energy-

INDEX AND GLOSSARY

converting system to the input stage of the same system, 72

fertilisation, 38, 41
see also **in vitro fertilisation**

fertiliser, 65

fibre, indigestible **complex carbohydrates** found in fruit and vegetables, 98, 106, 107, 108

fill (in road-building), soil brought in to raise ground level where ground is lower than road being built, 127

fillers, 88, 92
see also **composite materials**

first law of thermodynamics formal statement of the principle of the conservation of energy 30–31

fission (nuclear), process whereby a **neutron** causes a **nucleus** to split, with the release of energy and further neutrons which allow the **chain reaction** to proceed, 11–12

fission products, fragments from the nuclear **fission** process of **atoms** resulting from the **radioactive decay** of these fragments; several hundred fission products are produced in fuel while a reactor is operating, 11, 14, 16

flexible pavement (in road-building), **surfacing** and **road base** materials held together by bitumen, 129, 130

fluorescence, spontaneous re-emission of light from **atoms** that have been excited by **pumping**, 71–2

fluorocarbons, propellants used in some aerosols, polystyrene food containers and refrigerators, which are responsible for damaging the ozone layer and adding to the **greenhouse effect**, 37

follicle, structure within the ovary containing the egg and consisting of its surrounding nutritive and hormone-producing cells, 40, 41

follicle-stimulating hormone, 40

food containers, 90

food intolerance, adverse reactions to a specific food, not necessarily involving antibodies, 102–4

formation, surface of **subgrade** after completion of earthworks, 129

forward deduction, problem-solving by going from the starting state to a target state, 143

FOSD, full overtaking sight distance; distance required for drivers to overtake, while maintaining safety, 128–9

fossil fuels, fuels such as coal, oil and natural gas, produced by organic processes in past geological times, 34, 35, 36–7, 54

frequency (f), number of waves per second, 69

frozen eggs, 50

frozen embryos, 47

frozen pre-embryos, 47, 50

fructose (fruit sugar), **simple sugar**; soluble and sweet, 98

fuel cycle, 18–19

fuel element, form in which fuel is used in a nuclear reactor; the fuel element may be a group of small fuel rods (as in AGR or PWR reactors) or one large element, as in MAGNOX reactors, 12, 14

fuel element cans 14, 19

fusion (nuclear), fusing or joining together of **atoms**, with the release of energy; the energy source that powers the sun, 123

gametes, eggs and sperm, 38, 39

gamma radiation, 17

geiger counter, 18

genes, units in a **chromosome** which determine individual characteristics, e.g. eye and hair colour, 38, 39, 41

geothermal power, use of underground steam derived from hot dry rock or cooling lava/magma as source of heat **energy** for conversion into electricity, 112

glass fibre composites, 92

global climatic changes, 36–7

glucose, simple sugar; soluble and sweet, 98

goal (computer), objective of a **means-ends** computer **program**, 142

gravitational energy, form of **energy** that depends on the distance between objects, e.g. the moon and the Earth, 25, 26, 27, 28, 32

greenhouse effect, name given to the way various atmospheric pollutants (especially carbon dioxide) are altering the atmosphere so as to increase the average temperature of the Earth and thus change its climate, 37

half-life, time taken for half the **atoms** in a **radioactive** substance to **decay**, 16–17

hard water, slightly **alkaline** water, high in calcium and magnesium with low **plumbosolvency**, 64

heat exchanger, 13

heating systems, domestic, 31

Heimay, 112

hertz (Hz), unit of **frequency**; one cycle per second, 69

heuristics, broadly stated rules which have a better than average chance of finding a desired answer but which lack rigorous mathematical explanation, 139

high head hydro-electric system, hydro-electric system in which the water falls through a large

distance, e.g. down the side of a high valley, 26
high-temperature polymers, 95
Himalayas, 116
hologram, two-dimensional recording on a photographic plate that can **reconstruct** a very realistic three-dimensional image, 76–7
homopolymer, polymer made from a single type of **monomer**, 95
hormones, chemical messengers produced in specialised cells, transported via bloodstream to specific target organs whose activity they regulate, 40, 41, 42
hot spots, regions often well away from a plate edge where the Earth's mantle is hot enough to burn through the **lithosphere**, producing basaltic volcanoes, 117–18, 119
housekeeping commands, 136
hydro-electric power stations, 25–6, 32, 35–6
hydrogen ion, atom of hydrogen which has lost its negatively charged **electron**, leaving only a positive charge, 53
hypertension, high blood pressure, i.e. above 140/90 mmHg (millimetres of mercury), 105
hypocentre, location and depth of an earthquake, 120

illegal command, 136
implantation 41–2
in vitro fertilisation, fertilisation of egg(s) with sperm outside the body (literally 'in glass'), 38, 42, 43, 44–8, 51
in vivo fertilisation, fertilisation of egg(s) in the body, 38, 41
incidental pollution, 52
incoherent light, 69–70
incubation, maintaining in a constant environment, 45, 46

industrial discharge, 57
industrial lasers, 74–5
inert, chemically unreactive, 83
inference engine, part of an expert system which deduces facts from other facts and rules associating those facts, 144
infertility, 42–44
infra-red radiant energy, form of radiant energy, the wavelength of which is longer than light, that is converted directly into **thermal energy** when it is absorbed, 36
inherited disease diagnosis, 51
injection moulding, molten plastic forced into a mould at high pressure, 92, 93
inorganic substance, substance composed of **molecules** not based on carbon, 63
input device (computer), device which receives **instructions**, e.g. a keyboard or magnetic disk, 135
insemination, introducing sperm to medium containing egg, 46
instruction (computer), information telling **processing unit** where to find **data** and what to do with it, 134, 135, 136
insulation, thermal, 34, 35
intensity, power falling on a specified area, 70
interference, the overlapping of wave patterns which may produce a larger peak (constructive interference) or cancel each other out (destructive interference), 76
International Commission on Radiation Protection, 16
ionic bond, bond between positively and negatively charged ions, forming a new compound, 55
island arc, string of islands formed by underwater volcanic activity associated with the collision of two

ocean plates, 115–16
isotopes, forms of an **element** with the same chemical behaviour but differing numbers of **neutrons** in the **nucleus**, 10

kinetic energy, mechanical energy of motion, 11, 25, 26, 27, 28, 29, 32
knowledge-based system *see* **expert system**, 142
Krakatoa, 110

Lake District, 115, 119
laparoscopy, examination of abdominal organs by means of a laparoscope passed through an incision in the navel, 44
laser weapons, 78–80
lead pollution, 54, 63, 64
light waves/particles, 72
liming, 55–6
linoleic and linolenic acids, essential fatty acids required for health of skin and resistance to infection; deficiency cannot be made good from other dietary components, 97
lithosphere, outer rigid layers of the Earth's surface (**crust** and **upper mantle**), 113
Loila volano, 118
long sea outfall, coastal sewage pipeline long enough to prevent return of discharged sewage flow to the shore, 61
low head hydro-electric system, hydro-electric system in which water only falls through a small height, e.g. along a river or out of a low dam; tidal energy exploits a low head of water, 26
luteinising hormone, 41

MacAdam, John London, 123
MAGNOX reactors, 14, 19
Maiman, Theodore, 70
mantle, 113, 114, 120
see also **upper mantle**
means-ends analysis, program involving a search for a solution in order to

INDEX AND GLOSSARY

fulfill a specific objective, 142
mechanical energy, energy due to stresses and tensions (elastic energy) or due to the motion of an object (kinetic energy), 25, 29, 32
medical lasers, 75–6, 80–1
memory (computer), storage facility of a computer, 135
menstruation, 42
Metcalf, John, 123
micro-organisms, microscopic living organisms including bacteria, fungi, viruses and algae, 60
minerals, dietary, 100
MINI-MAX algorithm, 141
moderator, light material, such as water or graphite, forming part of the **core** of a 'thermal' nuclear reactor, used to slow down the fast **neutrons** from **fission** to increase the chance of further fission in ^{235}U, 12, 14
molecule, groups of **atoms** physically attached to each other, 85
monochromatic light, light of a single colour or **wavelength**, 67, 71
monomer, molecule capable of reacting with other molecules to form **polymers**, 82, 85, 87, 94
motility, sperm tail's active swimming movements, 39, 43, 45, 46

National Radiological Protection Board, 16
neural net, computers or computer **programs** which model the brain's **neurons**; computers designed to build up knowledge by learning or being trained, 146–9
neuron, basic structural and functional unit of the nervous system; nerve cell, 146
neutralisation, raising or lowering of pH to 7; resulting solution is neither **acid** nor **alkaline**, 55

neutron, neutral (no electric charge) particle of approximately equal mass to that of the **proton**; neutrons and protons together form the atomic **nucleus**, 10, 11, 12, 15, 17, 83
neutron poisons, 15
nitrate pollution, 63, 64–5
non-stick coatings, 90–1
nuclear energy, energy due to differences in mass between atomic **nuclei** given by Einstein's formula $E = mc^2$, 25
Nuclear Installations Inspectorate, 16
nucleus, closely packed group of **neutrons** and **protons** which form the central part of an **atom**; positive electrical charge is supplied by the protons, 8
nutrients, essential, 97–100
nutritional deficiencies, 100–101
nutritional supplements, vitamins, minerals or **essential fatty acids** which are taken in addition to a normal diet to ensure an adequate intake of essential nutrients, 102
nylon, 89

ocean floor spreading, formation of new oceanic **crust** from **upper mantle** associated with the moving apart of oceanic plate boundaries, 116, 119, 120
oceanic ridge, range of underwater mountains associated with strong and frequent earthquake or volcanic activity and formation of new oceanic **crust**, 113, 116
oestrogen, 40
opaque (to energy), absorbing **radiant energy**, 36
operating speed, overall speed of vehicle travelling under usual traffic and weather conditions, 127, 128
operators, statements (e.g. if, add) which instruct a computer to perform logical or arithmetical tasks, 137, 138
optical computing techniques, 81, 145–6
optical fibre, thin strand of glass that can transmit light with little loss over very long distances, 75, 81
organic chemistry, 86
organic substance, substance composed of **molecules** in which carbon is present, 63
orgasm, 39
output device (computer), device which displays information, e.g. a screen or printer, 135
ovulation, release of egg from ovary, 41
ovulation failure, 43
oxygen demand, oxygen required for metabolic activities of organisms, including those involved in breakdown of pollutants, 57–8
see also **biochemical oxygen demand**
oxygen saturation concentration, 58

parallax, changes in an image caused by moving the point of observation, e.g. from above to below, 76
pavement (in road-building), **surfacing, road base** and **sub-base**, 129
pavement materials, 130
perpetual motion machines, 32–3
persistent pollution, 52
PET (polyethyleneterephthalate), 89, 90, 92
pH scale, scale used to express the **hydrogen ion** concentration of a solution and thus the **acidity** (pH 0 to 7) and **alkalinity** (pH 7 to 14), 53
phase, position of a point on a wave, such as a peak or trough; two waves at their peak at the same position or time are 'in phase', 68, 76

photon, fundamental 'particle' of light, 72
photosynthesis, conversion by plants of carbon dioxide and water into oxygen and **carbohydrates** using the energy of sunlight, 111
pigment additives, 88
placenta, spongy structure formed by interlocking of fetal and maternal tissue in uterus; allows exchange of nutrients and gases; afterbirth, 41
plasma, 23
plasticisers, 88, 90
plate boundaries, 114, 115–18
plate tectonics, study of the movements of segments of the Earth's rigid **lithosphere**, 113–14
plumbosolvency, lead, e.g. from pipes, dissolving in water, 64
plutonium (Pu), 'man-made' metal which is **radioactive**; the **isotope** ^{239}Pu is readily fissile, forms the fuel of the **fast-breeder reactor** and is produced through **neutron** absorption in ^{238}U, 19, 20
polymer, very large **molecule** formed by joining lots of **monomer** molecules, 82–3, 85, 86, 88, 92
polymer biodegradation, 94–5
polymerisation, joining together of **monomer molecules** to form **polymers**, 86–7
polypropylene, 89
polystyrene, 89, 95
polythene, 85, 86, 87, 89, 90
polyunsaturated oils, vegetable oils (except palm oil and coconut oil), 97
Pompeii, 110
poor-quality river, river with polluted water containing no fish, 57
positive void coefficient, 15
power, rate of delivery of **energy**; energy per second, 57
precipitate, substance deposited as a solid from a solution, 64
pre-embryo, fertilised egg after fusion of parental genetic material; pre-implantation stages, 41, 42
donation, 50
freezing, 47, 50
research, 48
transfer, 47
Pressurised Water Reactor (OWR), 14, 19
problem-solving programs, 142–4
procedure, see **subroutine** 136
processing unit, part of a computer that performs arithmetical or logical tasks, 135
progesterone, 41, 42, 47
program (computer), complete list of **instructions** that a computer executes in sequence in order to perform a task, 135
programming (computer), 134–7, 142–3
PROLOG, 144
prompt neutrons, 15
pronuclei, male and female nuclear material within fertilised egg before combination of **chromosomes**, 41, 46
proteins, complex **organic molecules** composed of long folded chains of **amino acids**, 98–9
proton, particle with a mass about 2000 times that of an **electron** and with a positive charge equal to the negative charge on the electron found in the **nucleus** of an **element**, 10, 17, 83
PTFE (polytetrafluoroethylene), 89, 90, 91
pumped storage scheme, hydro-electric scheme that uses surplus output from power stations to replenish the water in its upper reservoir, 35–6
pumping (laser), supplying energy to a laser medium to excite it, 71
PVC (polyvinylchloride), 89–90, 93–4
pyrolysis, burning of plastics under controlled conditions, allowing original **monomers** to be recycled, 94

quantum mechanics, 72

radiant energy, only means by which **energy** can be transferred through a vacuum as electromagnetic waves similar to visible light, 25
radiation, release and spreading of gamma rays, x-rays or particles such as alpha and beta particles, 16, 17, 18, 21, 22
radiation dose, term normally used for the absorption of **radiation** by human tissue; measured in **sieverts**, 16, 18
radioactive waste, 18, 19, 20
radioactivity, spontaneous disintegration of **atoms**, often accompanied by the emission of **radiation**, 14, 16–17, 18
radon gas, radioactive gas emitted through **decay** of naturally occurring activity in rocks and soil, 22
reconstruction (holographic), re-illumination of holographic interference pattern with laser light to produce a lifelike three-dimensional image, 76
records (computer), units of related **data** stored together in a **database**, e.g. a person's name, address and telephone number could collectively constitute a record, 144
Red Sea, 118
refuelling of reactors, 13
renewable energy, power derived from **solar energy**, geothermal energy or tidal energy, 25, 32, 37
reproduction, normal, 38–42
resin, partially **polymerised**

INDEX AND GLOSSARY

compound, 93
retina, part of the eye sensitive to light, 75, 76
retinal welding, 75–6
rigid pavement (in road-building), **road base** and **surfacing** in form of concrete slab with cement binder, 129, 130
river pollution, 56–9
River Quality Survey (1985), 57
road base, distributes weight of traffic over **sub-base** or **formation**, 129, 130
road-building materials, 130–1
road design, 123–4, 127–8, 131, 132–3
road maintenance, 132–3
route location, 126–7
ruby laser, 70–1, 72
run of the river system, hydro-electric system which uses the **kinetic energy** of moving water to rotate a **turbine**, 25

safety fences, 132–3
saturated fats, fats with fully hydrogenated carbon chains; solid at room temperature and found mainly in animal products, e.g. butter and cheese, 97, 106
sea outfall, coastal sewage outlet, often discharging at low tide level, 60, 61, 63
seawater, sewage pollution, 59–61, 62–3
second law of thermodynamics, formal statement that **energy quality** always decreases, 31, 32
seismometer, instrument that detects ground movements as small as a thousandth of a millimetre, including volcanic tremor; records the arrival time and magnitude of the shock wave, 120, 121
Sellafield, 18, 19
sewage treatment, 57, 59–60, 61, 62–3, 65
semiconductor laser, miniaturised laser using semiconducting materials such as gallium arsenide, 78
seminal fluid (semen), sperm as well as activating and nutritive secretions of male genital glands, 39, 40, 45
Shannon, Claude, 137, 138, 139, 141
shear strength, thickness of material required to spread vehicle load on roads, 130
shield (in nuclear reactors), wall of material, often concrete, that encloses **radioactive** material or a nuclear reactor to prevent **radiation** harming operating personnel, 13
sievert (Sv), unit of **radiation dose** which takes account of the effectiveness of **radiation** in causing damage to human tissues, 18
simple sugars, soluble carbohydrates, usually sweet, 98
Sizewell enquiry, 14
slow drug release, 94
soft water, water which is low in calcium and magnesium, with high **plumbosolvency**, 64
solar energy, radiant energy from the sun, 25, 29, 32, 34, 36–7
speed of light (c), 69
sperm, 38, 39–40, 41, 43, 45, 46
disorders, 43, 46
injection, 50–1
preparation for IVF, 45–6
spoil tips (in road-building), sites used to take excess soil from **cut**, 127
SSD, stopping sight distance; distance travelled during thinking plus reaction time plus braking distance, 128
stabilisers, 88, 93–4
stimulated radiation, light that is generated when an excited **atom** is triggered by an incoming **photon**, 70, 72
Strategic Defence Initiative (SDI), 78–9
stretch blowing, plastic stretched in one direction and blown in another to align **polymer** chains in two directions, 92
stroke (and heart disease), 106
structure-property relationship, describes the ways in which changes in the arrangement and bonding of **atoms** within a substance alter its physical properties, 88, 89, 90, 95
styrene, monomer used in polystyrene production, 95
sub-base (in road-building), layer or layers of material supporting **road base**, 129, 130
subduction, the pulling down of a heavier plate beneath a lighter plate in the course of a collision, 115
sub-goal, state that must be reached first in order to reach a goal in a **means-ends program**, 142
subgrade (in road-building), upper part of soil supporting loads carried by **pavement**, 129, 130
subroutine (procedure), self-contained section of a **program** used to perform a specific task, as part of a larger program, 136
sucrose, table sugar; **carbohydrate**; a mixture of **glucose** and **fructose**; soluble and sweet, 98, 107, 108
surfacing (in road-building), provides safe and comfortable surface for traffic, 129, 130–1, 132
surgical swabs, 95
synapse, point at which information from one nerve cell can be transmitted to another, 149
synthetic fibres, 90, 95

Tangshan earthquake, 110
tap water pollution, 63–5
technical life, time taken for materials or construction of a road to wear out, 126
Telford, Thomas, 123
thermal energy, energy due to differences in temperature

between objects, 25, 29, 31, 32, 33, 35, 36–7
see also **geothermal power**
thermal insulation, 34, 35
thermodynamic efficiency, 33, 34
thermoplastic composites, 92–3
thermoplastics, plastics that can be melted and then moulded into any form, 91, 92
thermosetting composites, 93
thermosetting plastics, plastics in which the constituents are reacted together within a mould, and set to give a rigid, brittle plastic, 91–3
Three Mile Island, 14
thrombus, 105
tidal power, 28
TOKAMAK, 23
trace minerals, minerals required in tiny amounts for essential metabolic functions, 100
tracer chemicals, 62
traffic capacity, amount of traffic a road is designed to carry, 126
traffic demand calculation, 125
traffic loading, 130
transient (in nuclear reactors), unscheduled rise in power in a nuclear reactor, 15
transparent (to energy), not absorbing **radiant energy**, 36
Trésaguet, Pierre, 123
turbine, machine that converts flowing water or air into rotary mechanical motion, 13, 26, 28, 36
turbo-generator, combination of a **turbine** and a **dynamo** used in hydro-electric and wind-power devices, 26
Turing, Alan, 136

ultrasound directed follicle aspiration, egg collection using ultrasound probes, 44
upper mantle, inner layer of the **lithosphere**, 113
uranium-235, 8, 9, 11–12, 19
uranium-238, 11, 19, 20
uranium, natural, uranium as it is mined, containing 0.7% of ^{235}U and 99.3% of ^{238}U, 12, 19
uranium processing, 18, 19–20
urethra, genital tract which carries semen through the penis; carries urine from bladder in both sexes, 39

vacuum forming, molten plastic drawn into a mould using a vacuum, 92
vitamin deficiencies, 100
vitamins, 99–100
volcanic soils, 112
volcanic winter, global temperature drop lasting for many months, caused by volcanic ash or gases screening the Earth from the sun's rays, 110
volcano monitoring, 121

water authority licences, 59
wave power, 27–8, 32, 37
wave velocity (c), **frequency** (f) multiplied by **wavelength** (γ); speed of the wave motion, 69
wavelength (γ), distance between two successive peaks or troughs of a wave, 68
welding, laser, 74
wind power, 37
wind turbines, 27, 32
WISARD, 149
wound sutures, 94

X chromosome, 39, 46

Y chromosome, 39, 46

zinc pollution, 54